STRATIGRAPHY of the SUBSURFACE LOWER and MIDDLE DEVONIAN of NEW YORK, PENNSYLVANIA, OHIO and ONTARIO

by

Lawrence V. Rickard

Geological Survey

NEW YORK STATE MUSEUM
MAP AND CHART SERIES 39

1989

The University of the State of New York
The State Education Department
Albany, New York 12230

ISSN 0097-3793
ISBN 1-55557-181-6

CONTENTS

ILLUSTRATIONS

Plates [Isopach-Lithofacies Maps, in box]

Paleogeologic Map

STRATIGRAPHY of the SUBSURFACE LOWER and MIDDLE DEVONIAN of NEW YORK, PENNSYLVANIA, OHIO and ONTARIO

by

Lawrence V. Rickard

ABSTRACT

Fourteen cross sections, seventeen isopach-lithofacies maps, and one paleogeologic map are used to depict the stratigraphic relationships and subsurface distribution of nineteen rock units of the Lower through Middle Devonian in New York, Pennsylvania, Ohio and Ontario. This report is based on a study of sample logs and gamma-ray logs of over 1000 wells. New correlations resulting from this study are proposed. These include new information on the relationship of the Onondaga Limestone and the Hamilton Group of New York to the Selinsgrove Limestone and Mahantango Formation, respectively, of Pennsylvania.

INTRODUCTION

This report is the culmination of a study of 676 sample logs and 545 gamma-ray logs from over 1000 wells distributed throughout New York (624 wells), Pennsylvania (317 wells), north-eastern Ohio (42 wells), and southern Ontario (25 wells) (Figure 1). The work was nearly completed by 1985 and data from wells drilled later have not been incorporated into the study. The stratigraphic interval extends from the Upper Silurian Salina Group through the Lower Devonian Helderberg and Tristates Groups, the Middle Devonian Onondaga Limestone, Hamilton Group and Tully Limestone, and into the Upper Devonian Genesee and Sonyea Groups (Figure 2). Stratigraphic cross sections and isopach maps were drawn for all formations or groups and for some subdivisions thereof. These maps and sections display the stratigraphic relationships and geographic distribution of these units in the subsurface throughout the study area. New correlations resulting from this research are included in Figure 3.

Twenty-five cross sections covering the study area (Plates 1-22) have been drawn at a vertical scale of 1 inch = 100 feet. The datum plane for all sections is the middle ash bed of the Tioga Bentonite [i.e., Tioga Bed B as denoted herein and by Rickard (1984)]. On many cross sections the Onondaga and Tully Formations are redrawn at double the scale of the main section to better illustrate details of their subdivisions.

Seventeen isopach and lithofacies maps (Plates 23-39) show the thickness, geographic distribution and lithology of the stratigraphic units. These maps originally were drawn at 1:500,000 but were reduced to 1:750,000 for publication. A paleogeologic map of the study area at the base of the Middle Devonian Onondaga Limestone (Plate 40) shows the rock unit that the drill will penetrate beneath the Onondaga.

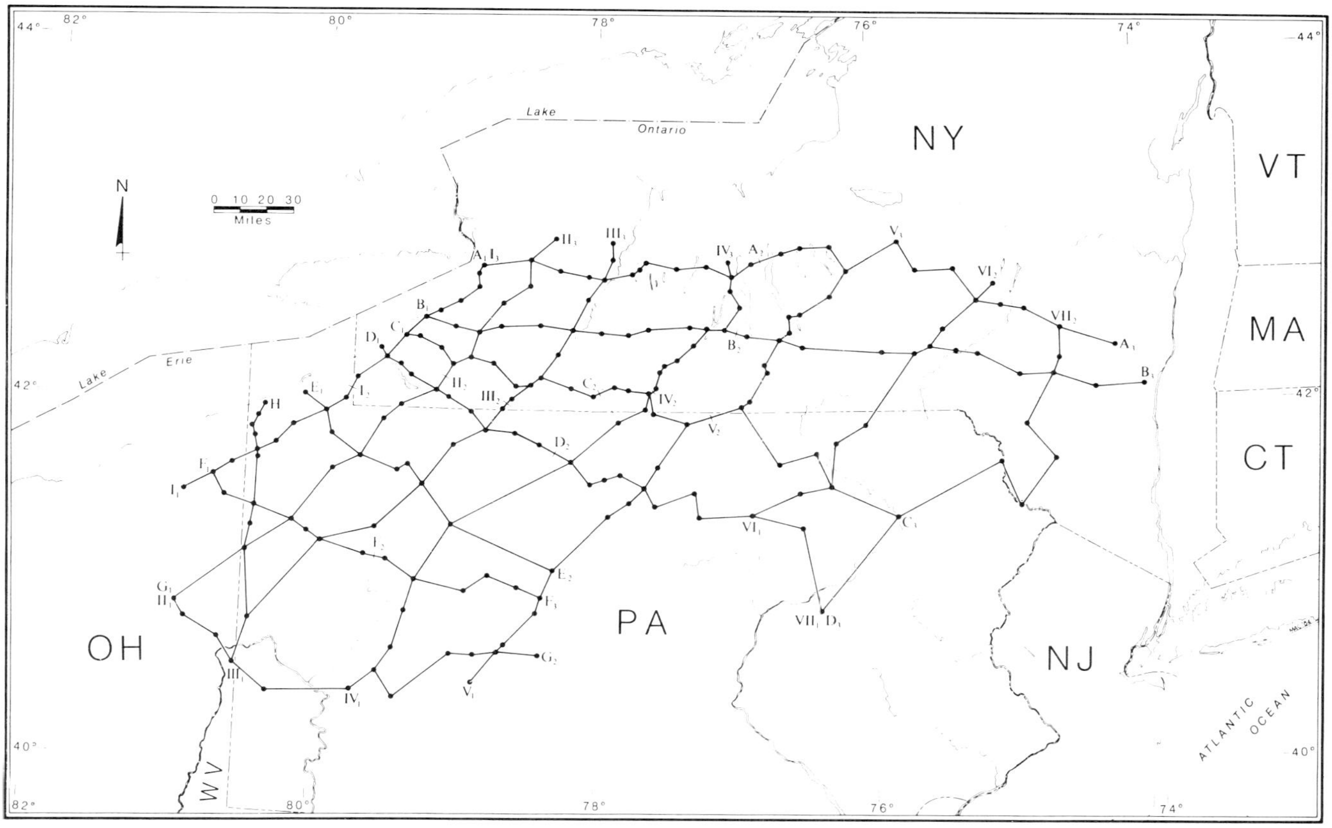

Figure 1. Location of area investigated

RONDOUT AND BASS ISLANDS FORMATIONS

Rickard (1962) reviewed the different uses of the name "Rondout" in central and eastern New York. Presently, the name is applied to limestone and dolostone strata subjacent to the Manlius Limestone throughout New York State. The nature and subdivision of the Rondout of eastern and southeastern New York differs from that of central New York. In the vicinity of Kingston, the Rondout Formation consists of four members. In ascending order these are: the Wilbur Limestone, Rosendale Dolostone, Glasco Limestone, and Whiteport Dolostone. The members compose an interval of alternating beds of fossiliferous limestone and barren dolostone. Harper (1969) divided the Rondout in central and western New York into three members: the Cobleskill Limestone, the Akron Dolostone, and the Chrysler Dolostone. The Chrysler overlies both of the other members (Belak, 1980; Ciurca, 1982). Fossils found in the limestone members of the Rondout (Cobleskill, Wilbur, and Glasco members) indicate a very Late Silurian age for these parts of the formation. The upper part of the Rondout, the Whiteport and upper Chrysler dolostones, lack diagnostic fossils and may be either Late Silurian or Early Devonian in age.

The Rondout Formation of New York can be traced through the subsurface into northern Pennsylvania where it is equivalent to part of the lower Keyser Formation (Figure 3). To the west, the Rondout grades into the Bass Islands Formation of Ohio, Ontario, and Michigan. The present distribution of the Rondout-Bass Islands is shown in Plate 23. The formation is thickest in south-central New York where it exceeds 100 feet. Erosion removed it throughout most of Erie, Wyoming, Genesee, and western Livingston Counties, New York, prior to deposition of the Onondaga Limestone. The Rondout is also missing in much of Ontario and northern Yates Counties for the same reason.

The geographic distribution of the Rondout is of some interest to the petroleum industry because of recent discoveries of oil in the basal member of the formation in Chautauqua County. The Akron Dolostone member (equivalent to the Cobleskill Limestone in central and eastern New York) has been fractured by faulting along a narrow SW-NE trend that extends across the county. Some wells that penetrate this fractured zone are commercial producers (Copley and Gill, 1983).

- Upper Devonian
  - Sonyea Group
    - Cashaqua Shale
    - Middlesex Shale
  - Genesee Group
    - West River Shale
    - Penn Yan Shale
    - Geneseo Shale
- **Middle Devonian**
    - **Tully Limestone**
  - **Hamilton Group**
    - **Moscow Formation**
    - **Ludlowville Formation**
    - **Skaneateles Formation**
    - **Marcellus Formation**
  - **Onondaga Formation**
    - **Seneca Limestone**
    - **Moorehouse Limestone**
    - **Nedrow Limestone**
    - **Clarence and Edgecliff Limestones**
- **Lower Devonian**
  - **Tristates Group**
    - **Schoharie and Bois Blanc Limestones**
    - **Carlisle Center Siltstone**
    - **Esopus Shale**
    - **Oriskany Sandstone and Glenerie Limestone**
  - **Helderberg Group**
    - **Port Ewen Limestone**
    - **Alsen Limestone**
    - **Becraft Limestone**
    - **New Scotland Limestone**
    - **Kalkberg Limestone**
    - **Coeymans Limestone**
    - **Manlius Limestone**
- **Upper Silurian**
    - **Rondout and Bass Islands Dolostones**
  - Salina Group
    - Bertie Dolostone
    - Camillus Shale
    - Syracuse Formation
    - Vernon Formation
  - Lockport Group

Figure 2. Stratigraphic column. Units discussed are in boldface.

HELDERBERG GROUP

The stratigraphy and paleontology of the Helderberg Group were described by Rickard (1962) and Epstein and others (1967). Subsequently, a series of paleoenvironmental analyses were initiated by Laporte (1967, 1969) and continued by Anderson (1967, 1971, 1972), Harper (1969), Head (1969), Epstein (1970), and Arif (1973). The Helderberg Group consists of a sequence of limestone formations of different lithologies and fossil content. In ascending order, these are the Manlius, Coeymans, Kalkberg, New Scotland, Becraft, Alsen, and Port Ewen Formations. Dunn and Rickard (1961) subdivided the Kalkberg Limestone in the Hudson Valley.

The Manlius Formation consists of fine-grained limestone beds with faunas and sedimentary structures indicative of deposition in supratidal, intertidal, and shallow subtidal environments. The overlying Coeymans Formation is coarse-grained, with coral bioherms or reefs, and was deposited in a high energy, subtidal environment. The Kalkberg and New Scotland Formations are the most fossiliferous units of the Helderberg Group. The former consists of medium-grained cherty limestones whereas the latter is argillaceous. They represent deposition in progressively deeper subtidal environments. The higher Helderberg formations are similar to the lower ones, that is, the Becraft is like the Coeymans, the Alsen is like the Kalkberg, and the Port Ewen is like the New Scotland.

The formations of the Helderberg Group can be recognized in the subsurface through the use of gamma-ray and sample logs (Waldman, 1979). The group provides an excellent subject for using gamma-ray logs to correlate different carbonate facies because gamma-ray patterns persist laterally from one facies into another without much modification and permit the correlation of different carbonate rocks with dissimilar faunas.

This study differentiated the individual formations of the Helderberg Group. They show so little variation in thickness that a separate isopach map for each unit is not warranted. Plate 24 is an isopach map of the entire group in New York and its equivalents in adjacent areas. The group is beveled out to the west below by an unconformity and is overlain either by the Tristates Group or by the Onondaga Limestone where the former is absent. The highest Helderbergian formation present locally is indicated on Plate 24. The group reaches a maximum thickness of over 400 feet in northeastern Pennsylvania and is absent in western New York and northwestern Pennsylvania.

TRISTATES GROUP

Rickard (1975) proposed the name "Tristates Group" for those rock units including the Port Jervis Formation through the Schoharie Formation. The type locality forms the top and west flank of Trilobite Mountain, north of the village of Tristates in the Port Jervis Quadrangle. The group includes, in ascending order, the Port Jervis Limestone, Glenerie Limestone, Oriskany Sandstone, Esopus Shale, Carlisle Center Siltstone, and the Schoharie Formation. The Port Jervis Limestone is present only in the vicinity of Port Jervis and, therefore, was not included in this study. In eastern Pennsylvania, the Schoharie Formation includes the Carlisle Center Siltstone (Epstein, 1984). The most important papers on the rocks of this group in New York are by Laskowski (1956), Johnsen (1957), Boucot (1959), Johnsen and Southard (1962), and Boucot and others (1970).

Oriskany and Glenerie Formations

Plate 25 is an isopach map of the Oriskany Sandstone and its equivalent in southeastern New York, the Glenerie Limestone. The Oriskany is a quartz arenite with calcareous cement. It is a deposit of a high-energy, shallow subtidal environment with a fauna characterized by large brachiopods, particularly spiriferids and terebratulids. The Glenerie Limestone is a moderately fossiliferous, highly siliceous limestone with abundant black chert.

The Oriskany-Glenerie isopach map (Plate 25) indicates several areas in which the sandstone exceeds 100 feet in thickness. These are located in central and northwestern Pennsylvania and in northeastern Pennsylvania and southeastern New York. Many outliers of Oriskany occur in New York and Pennsylvania; a few appear in northeastern Ohio and one is known in Ontario. The Glenerie Limestone exceeds 150 feet in thickness near Port Jervis in southeastern New York. The present distribution of the Oriskany-Glenerie is the result of modification by erosion during the Early Devonian. Except in eastern New York and northeastern Pennsylvania where it is overlain by the Esopus Formation, the top of the Oriskany-Glenerie interval is a major unconformity. In western New York and northwestern Pennsylvania, the Oriskany Sandstone has been separated from reworked sands (=Springvale and Sylvania) at the base of the Bois Blanc Formation and/or Onondaga Limestone. These younger, reworked sands have been confused with the Oriskany in the past. The Oriskany Sandstone has medium to coarse quartz sandstones whereas the younger Springvale sediments are typically silt-sized and argillaceous. Thus, Plate 25 is an attempt to draw an isopach map of the "true" Oriskany and does not include younger reworked sands. This map therefore differs from earlier ones (Cate, 1961; Kreidler and others, 1972; Heyman,

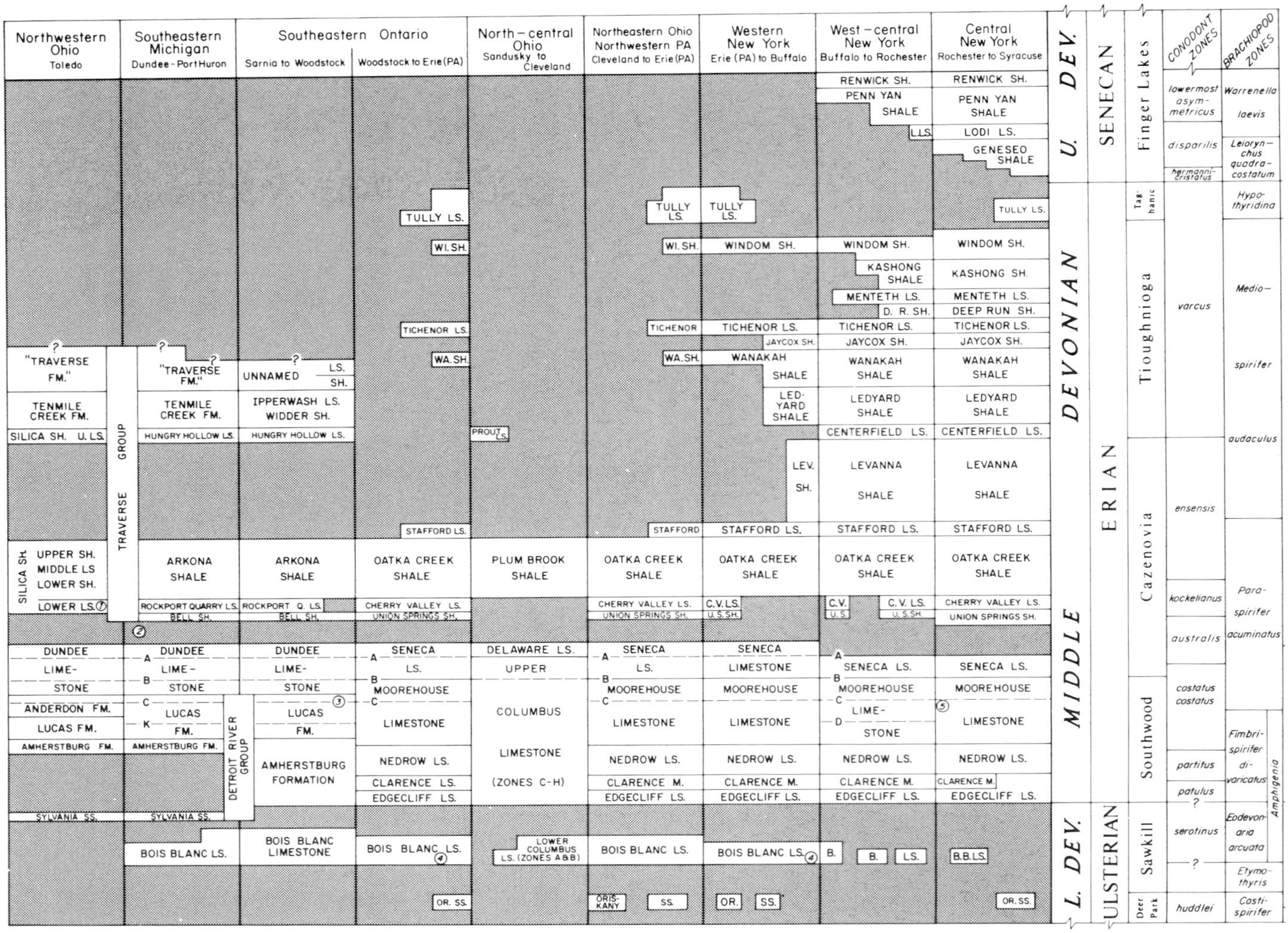

① So called "blue bed" at base of Silica Shale.

② Position of Rogers City Limestone in northern Michigan.

③ Position of so called "Columbus" Limestone near Ingersoll with *Prismatophyllum conjunctum*.

④ Includes sporadic occurrences of Springvale Sandstone at base.

⑤ *Acinophyllum-Eridophyllum* bed and overlying occurrence of *Prismatophyllum conjunctum*.

Figure 3. Correlation of the Lower and Middle Devonian.

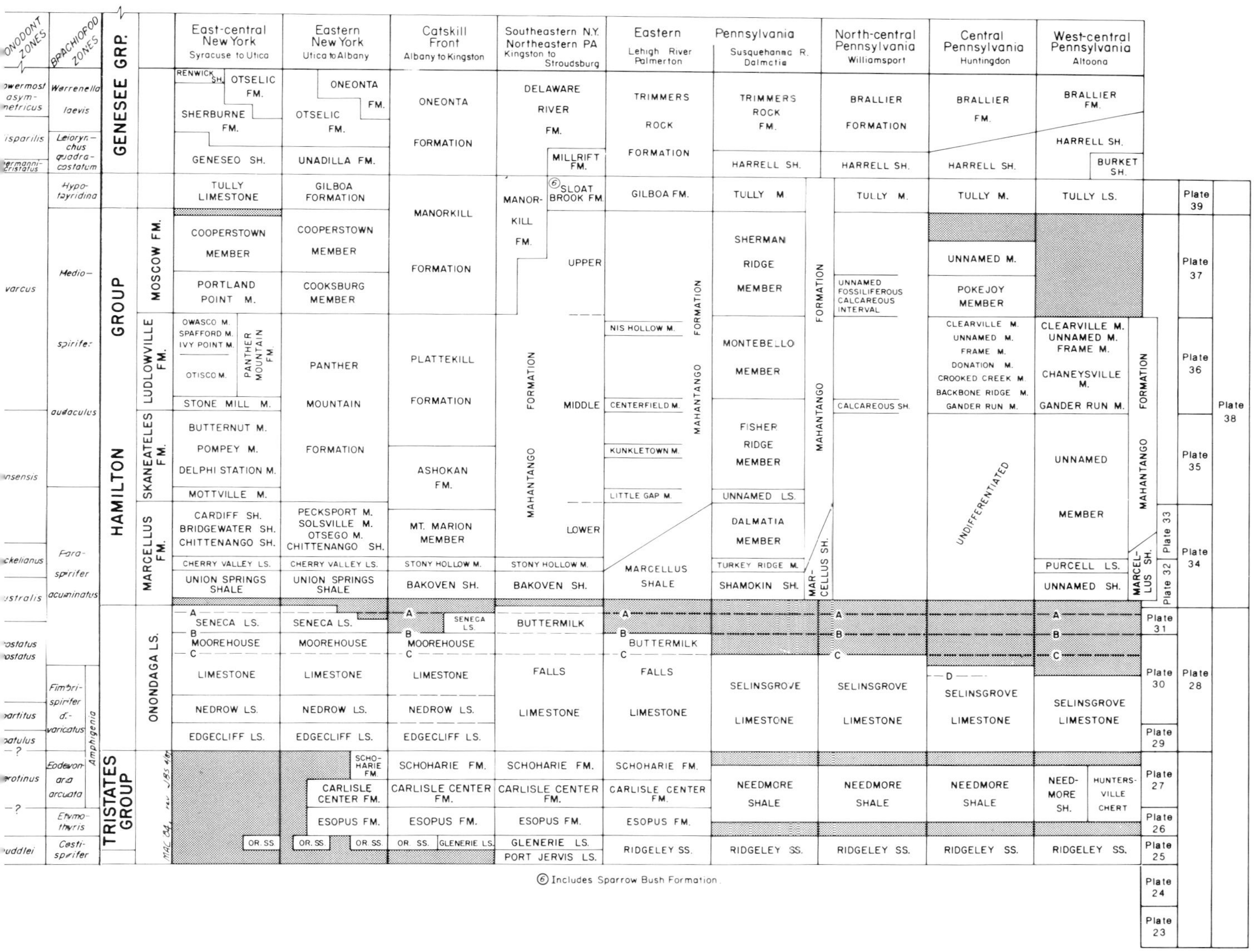

CONODONT ZONES
BRACHIOPOD ZONES
GENESEE GRP.
HAMILTON GROUP
ONONDAGA LS.
TRISTATES GROUP
MOSCOW FM.
LUDLOWVILLE FM.
SKANEATELES FM.
MARCELLUS FM.
East-central New York
Syracuse to Utica
Eastern New York
Utica to Albany
Catskill Front
Albany to Kingston
Southeastern N.Y. Northeastern PA
Kingston to Stroudsburg
Eastern
Lehigh River Palmerton
Pennsylvania
Susquehanna R. Dalmatia
North-central Pennsylvania
Williamsport
Central Pennsylvania
Huntingdon
West-central Pennsylvania
Altoona
Lowermost asymmetricus
disparilis
hermanni-cristatus
varcus
ensensis
kockelianus
australis
costatus costatus
partitus
patulus
serotinus
dehiscens
Warrenella laevis
Leiorhynchus quadracostatum
Hypothyridina
Mediospirifer audaculus
Paraspirifer acuminatus
Fimbrispirifer cf. varicatus
Amphigenia
Eodevonaria arcuata
Etymothyris
Costispirifer
RENWICK SH.
OTSELIC FM.
SHERBURNE FM.
GENESEO SH.
ONEONTA FM.
OTSELIC FM.
UNADILLA FM.
ONEONTA FORMATION
DELAWARE RIVER FM.
MILLRIFT FM.
TRIMMERS ROCK FORMATION
TRIMMERS ROCK FM.
HARRELL SH.
BRALLIER FORMATION
HARRELL SH.
BRALLIER FM.
HARRELL SH.
BRALLIER FM.
HARRELL SH.
BURKET SH.
TULLY LIMESTONE
GILBOA FORMATION
MANORKILL FORMATION
MANORKILL FM.
⑥SLOAT BROOK FM.
GILBOA FM.
TULLY M.
TULLY M.
TULLY M.
TULLY LS.
COOPERSTOWN MEMBER
PORTLAND POINT M.
COOPERSTOWN MEMBER
COOKSBURG MEMBER
UPPER
SHERMAN RIDGE MEMBER
MAHANTANGO FORMATION
UNNAMED FOSSILIFEROUS CALCAREOUS INTERVAL
UNNAMED M.
POKEJOY MEMBER
OWASCO M.
SPAFFORD M.
IVY POINT M.
OTISCO M.
PANTHER MOUNTAIN FM.
STONE MILL M.
PANTHER MOUNTAIN FORMATION
PLATTEKILL FORMATION
MAHANTANGO FORMATION
MIDDLE
NIS HOLLOW M.
CENTERFIELD M.
KUNKLETOWN M.
LITTLE GAP M.
MONTEBELLO MEMBER
FISHER RIDGE MEMBER
UNNAMED LS.
CALCAREOUS SH.
CLEARVILLE M.
UNNAMED M.
FRAME M.
DONATION M.
CROOKED CREEK M.
BACKBONE RIDGE M.
GANDER RUN M.
CLEARVILLE M.
UNNAMED M.
FRAME M.
CHANEYSVILLE M.
GANDER RUN M.
BUTTERNUT M.
POMPEY M.
DELPHI STATION M.
MOTTVILLE M.
ASHOKAN FM.
UNDIFFERENTIATED
UNNAMED MEMBER
CARDIFF SH.
BRIDGEWATER SH.
CHITTENANGO SH.
PECKSPORT M.
SOLSVILLE M.
OTSEGO M.
CHITTENANGO SH.
MT. MARION MEMBER
LOWER
DALMATIA MEMBER
MARCELLUS SH.
CHERRY VALLEY LS.
CHERRY VALLEY LS.
STONY HOLLOW M.
STONY HOLLOW M.
TURKEY RIDGE M.
PURCELL LS.
UNION SPRINGS SHALE
UNION SPRINGS SHALE
BAKOVEN SH.
BAKOVEN SH.
MARCELLUS SHALE
SHAMOKIN SH.
UNNAMED SH.
MARCELLUS SH.
A
SENECA LS.
B
MOOREHOUSE
C
LIMESTONE
SENECA LS.
MOOREHOUSE LIMESTONE
SENECA LS.
BUTTERMILK FALLS LIMESTONE
BUTTERMILK FALLS LIMESTONE
D
SELINSGROVE LIMESTONE
SELINSGROVE LIMESTONE
SELINSGROVE LIMESTONE
SELINSGROVE LIMESTONE
NEDROW LS.
EDGECLIFF LS.
NEDROW LS.
EDGECLIFF LS.
NEDROW LS.
EDGECLIFF LS.
SCHOHARIE FM.
SCHOHARIE FM.
SCHOHARIE FM.
SCHOHARIE FM.
CARLISLE CENTER FM.
CARLISLE CENTER FM.
CARLISLE CENTER FM.
CARLISLE CENTER FM.
NEEDMORE SHALE
NEEDMORE SHALE
NEEDMORE SHALE
NEEDMORE SH.
HUNTERSVILLE CHERT
ESOPUS FM.
ESOPUS FM.
ESOPUS FM.
ESOPUS FM.
OR. SS
OR. SS
OR. SS
OR. SS.
GLENERIE LS.
GLENERIE LS.
PORT JERVIS LS.
RIDGELEY SS.
RIDGELEY SS.
RIDGELEY SS.
RIDGELEY SS.
RIDGELEY SS.
Plate 39
Plate 37
Plate 36
Plate 35
Plate 34
Plate 33
Plate 32
Plate 31
Plate 30
Plate 29
Plate 28
Plate 27
Plate 26
Plate 25
Plate 24
Plate 23
Plate 38
⑥ Includes Sparrow Bush Formation.

1977; Abel and Heyman, 1981; Diecchio and others, 1984).

Esopus Formation

The dark gray siliceous shales and cherts that form the Esopus Formation are nearly barren of fossils. As shown on Plate 26, this formation is restricted to eastern New York and northeastern Pennsylvania. A maximum thickness of 414 feet occurs in the Hess well near the Delaware River in Pike County, Pennsylvania. Its present distribution suggests that the source area of the Esopus lies east of Port Jervis.

Carlisle Center, Schoharie and Bois Blanc Formations

These three formations are closely related in age and distribution. The Schoharie Formation and underlying Carlisle Center Formation occupy an area in eastern New York and Pennsylvania that is separated from the Bois Blanc Limestone in western New York, western Pennsylvania, Ohio and Ontario (Plate 27.) The Carlisle Center Formation consists of calcareous siltstone and silty shale and is overlain by calcareous siltstone and arenaceous limestone assigned to the Schoharie. Oliver (1966, 1967) described the Bois Blanc Limestone of New York, and Rickard (1984, 1985) discussed its occurrence in Pennsylvania, Ohio and Ontario.

The Carlisle Center-Schoharie interval is over 400 feet thick in northeastern Pennsylvania (Plate 27) where it is known as the Needmore Shale. These formations thin westwardly and pinch out in central Pennnsylvania and New York. Scattered erosional outliers are present in central and west-central New York and north-central Pennsylvania. The Needmore Shale is replaced by the Huntersville Chert in Western Pennsylvania which, in turn, is transitional into the Bois, Blanc Limestone in Ohio and Ontario. The Bois Blanc thickens to more that 200 feet in eastern Ohio. The Bois Blanc and Huntersville Chert are entirely older than the overlying Onondaga Limestone. Over much of the area, the contact between these formations and the Onondaga is an unconformity.

PRE-ONONDAGA PALEOGEOLOGY

The paleogeologic map of the study area immediately preceding deposition of the Onondaga Limestone is particularly interesting. Plate 40 shows the formation the drill is likely to penetrate beneath the Onondaga. These units include the Upper Silurian Bertie Dolostone of the Salina Group and the Upper Silurian-Lower Devonian Rondout Formation in north-central New York, various formations of the Lower Devonian Helderberg Group in east-central New York, and the Oriskany Sandstone, Carlisle Center and Schoharie Formations of the Tristates Group in eastern New York. In Pennsylvania and Ohio, they include the Needmore Shale, Huntersville Chert and Bois Blanc Limestone.

This paleogeology illustrates a regional angular unconformity at the base of the Onondaga Limestone. The Onondaga overlies successively older units on a line from north-central Pennsylvania (Oriskany Sandstone) to Erie County, New York (Bertie Dolostone). Presence of outliners of Bois Blanc Limestone on the Bertie Dolostone and below the sub-Onondaga unconformity indicates significant, regional uplift in the late Early Devonian along a southwesterly trend in western New York and adjacent Pennsylvania.

ONONDAGA LIMESTONE

The Middle Devonian Onondaga Limestone and its coral reefs have become well-known through the work of W.A. Oliver, Jr. (1954, 1956a, 1956b, 1960, 1963, 1966, 1967). Ozol (1963) and Lindholm (1967) described Onondaga petrology in greater detail while Rickard (1984, 1985) described the distribution and correlation of the Onondaga and its equivalents in the Lake Erie region.

Oliver and Ozol subdivided the Onondaga in New York State into five members. In ascending order these are the Edgecliff Limestone and Clarence Chert in the lower part of the formation, the overlying Nedrow and Moorehouse Limestones, and the Seneca Limestone at the top of the formation. Ash beds assigned to the Tioga Bentonite occur at four different levels within the Onondaga. Rickard (1984) described their relationships and distribution and gave them the letters A, B, C, and D, in descending order. These are shown in the cross sections included in this report. Tioga Ash Bed B is the datum plane in these sections.

Plate 28 is an isopach map of the Onondaga Limestone and its equivalents. The formation is thinnest along an axis extending northeastwardly across central Pennsylvania into central New York. The Onondaga is less than 20 feet thick along much of this axis, and it is known as the Selinsgrove Limestone. It thickens both northwestwardly and southeastwardly. The Onondaga equivalent in eastern Pennsylvania, the Buttermilk Falls Limestone, exceeds 200 feet in thickness. The Upper Columbus and Delaware Limestones of Ohio and the Onondaga of western New York and northwestern Pennsylvania are over 200 feet thick in several areas (Plate 28). Onondaga equivalents in Ontario, the Amherstburg and Dundee Limestones, thicken westwardly to over 400 feet.

Clarence and Edgecliff Members

An isopach map of the lowest member of the Onondaga Formation, the Edgecliff Limestone, and its equivalent in western New York, the Clarence Chert, is shown on Plate 29. The northeastwardly trending axis is evident on this map. Along this axis, the Clarence-Edgecliff Member is less than 10 feet thick in most places, but it is three to six times thicker both to the northwest and southeast. The small number of wells in eastern New York and eastern Pennsylvania limits our understanding of the member in this area.

The Edgecliff Member of the Onondaga has prominent coral reefs. In the subsurface, some of these reefs have been important sources of natural gas. Plate 29 shows the distribution of reefs known along the outcrop and in the subsurface. A relationship of reef distribution to the isopachs of the member is not apparent.

Nedrow and Moorehouse Members

Although the axis can be discerned on the isopach map of the Nedrow and Moorehouse Members of the Onondaga, it is not strongly developed. These members of the Onondaga are commonly less than 10 feet thick along the axis and may be absent in several areas. As the other units of the Onondaga, these members thicken both northwestwardly and southeastwardly of the axis. They are more than 100 feet thick in eastern New York and eastern Pennsylvania and are over 120 feet thick in Ohio, northwestern Pennsylvania and Ontario. They thicken westwardly in Ontario and exceed 300 feet.

Seneca Member

The presence of an unconformity at the top of the Onondaga Formation is best illustrated by the isopach pattern of the Seneca Member. This member is absent throughout a large area in central and western Pennsylvania (Plate 31) where the shales of the overlying Hamilton Group rest on the Moorehouse Limestone (and even on the Edgecliff in a few small areas, see Plate 30). The Seneca Member reaches 40 feet in thickness in eastern Pennsylvania and in northeastern Ohio. It may exceed 50 feet beneath Lake Erie.

Tioga Ash Bed B occurs at the contact between the Seneca and the overlying Moorehouse, and Ash Bed A occurs at or near the top of the Seneca. This relationship prevails along the outcrop belt of the Onondaga in New York and for a distance into the subsurface. However, the Seneca pinches out toward the axis and Ash Beds A and B merge into one composite ash bed. Because the upperpart of the the Moorehouse also pinches out, Ash Beds A and B eventually merge with the Ash Bed C. This amalgamation of ash beds can be seen in several of the cross sections in this report. For example, along the central axis of the basin where the Onondaga consists of the Edgecliff, Nedrow and lower Moorehouse Members, the three ash beds of the Tioga have merged into the single tripartite bed first described by Fettke in 1949 (Ebright and others, 1949). The presence of more than one ash bed and their amalgamation into the original Tioga was first described by Rickard (1984).

HAMILTON GROUP

A wedge-shaped mass of siliciclastics, the Hamilton Group consists of shale, siltstone and sandstone beds of marine and non-marine origin. Several thin but extensive limestone beds can be used to subdivide the group into formations (Cooper, 1930, 1933). In western New York, the group consists of fossiliferous marine gray or black shale with interbedded thin limestones. These rocks grade eastward into arenaceous shale, siltstone and sandstone beds of shallow-water marine origin (Cooper, 1933). Beneath the Catskill Mountains much of the Hamilton Group is nonmarine red and green shale and cross-bedded sandstone with plant and fish remains (Rickard, 1975).

These rocks have been the subject of many reports, but modern stratigraphic classification and correlation has developed from the classic work of G. A. Cooper (1930, 1933, 1934, 1941) and Cooper and Warthin, (1941, 1942). Additional features of the Hamilton Group have been described by Smith (1935), Oliver (1951), Rickard (1952), and McCave (1968, 1973). Paleoenvironmental studies have been completed by Fernow (1961), Grasso (1970), Mazullo (1973), and Way (1972). Considerable new information on Hamilton Group stratigraphy, paleontology, and paleoecology has been accomplished by Baird (1979,1981), Baird and Brett (1981, 1983, 1986), Gray (1984), Brett (1974, 1986), Brett and Baird (1982,1985) and Brett and others (1986). A redefinition of the Moscow Formation is an important contribution by Baird (1979). He redefined the base of this formation as the base of the Tichenor Limestone rather than the base of the younger Menteth Limestone.

The Hamilton Group of central and western New York State is composed of four formations. These are, in ascending order, the Marcellus, Skaneateles, Ludlowville and Moscow. Each has a thin limestone member at or near its base (Figure 3). Several of these limestones (e.g., Mottville, Centerfield) grade into shales and sandstones in east-central New York. Along the eastern outcrop belt, strata equivalent to the upper part of the Marcellus Formations as well as the Skaneateles and Ludlowville Formations have not been differentiated and are termed the "Panther Mountain Formation." Non-marine Hamilton strata are known as the Plattekill and Manorkill Formations (formerly grouped as the "Kiskatom Formation").

Plates 34 through 37 are isopach maps for the four formations of the Hamilton Group in western and central New York. Two additional isopach maps (Plates 32 and 33) show the subdivisions of the Marcellus Formation: 1) the Cherry Valley Member and underlying Union Spring Member and 2) the Oatka Creek Member and its eastward equivalents and reveal interesting details of lower Hamilton stratigraphy and geological history.

The Hamilton Group in Pennsylvania consists of two formations, the Marcellus and overlying Mahantango (Figure 3). These are strictly lithologically-defined units; the former is black or very dark gray shale whereas the latter comprises gray shale, siltstone and sandstone. Post-Hamilton Group strata that are equivalent to the Tully Limestone of New York overlie the Mahantango Formation; however, in some parts of Pennsylvania they have been included in the Mahantango and referred to the Hamilton Group (Faill and others, 1978; Berg and others, 1983).

Rickard (1984, 1985) described the correlation of Hamilton Group of New York into Ontario, Ohio, and Michigan. Briefly summarized, the Hamilton equivalents of these areas consist of Marcellus-correlative shales and thin limestones which are overlain locally Ludlowville and Moscow(?) equivalents.

The isopach map of the Hamilton Group (Plate 38) shows a steady west-to-east increase in the thickness of the group across New York State, in contrast to the many deviations from this pattern in the isopachs of the subdivisions of the group (Plates 32-37). No evidence exists for any thinning of the group in eastern New York. The Hamilton Group is 200 feet thick beneath western Chautauqua County and 500 feet thick beneath the Genesee River Valley at Wellsville. The rate of thickness increase is greatest eastward of Bath (central Steuben County), and the Hamilton Group is 1000 feet thick beneath Watkins Glen, 1200 feet at Ithaca and 1900 feet at Binghamton. Thickness continuously increases eastward to the Catskill Front where the Hamilton Group is 3250 feet thick.

Marcellus Formation

This name is applied in New York State to strata between the top of the Onondaga Limestone and the base of the Stafford- Mottville Members of the overlying Skaneateles Formation (Figure 3). The Marcellus Formation is a lithostratigraphic unit defined by key beds. In Pennsylvania and further southwest, the name "Marcellus" is applied to the black shale unit in the lower part of the Hamilton Group. Unfortunately, the upper boundary of the Marcellus is drawn in Pennsylvania at significantly different stratigraphic levels that depend upon the upward extent of the black shales beneath the overlying Mahantango Formation. In this report, the term "Marcellus Formation" refers to the unit as it is defined in New York; the designation "Marcellus Shale" refers to the unit as it is defined in Pennsylvania and southward. In the isopach map of the Marcellus Formation (Plate 34), the New York definition is used in Pennsylvania because the Stafford-Mottville interval that forms the uppermost Marcellus Formation in New York has been identified in Pennsylvania in the gamma-ray logs of wells and in certain outcrop sections.

The Marcellus Formation is thinnest in Cattaraugus County where it is less than 25 feet thick (Plate 34). The formation is less than 150 feet thick throughout much of western New York and western Pennsylvania. It thickens rapidly eastward of Watkins Glen in central New York (200 feet) to 650 feet at Binghamton. The rate of thickening decreases across Broome and Delaware Counties. Beneath eastern Delaware County the formation again begins to increase rapidly in thickness and thickens from 900 feet to 1800 feet along the Catskill Front. The Shaller well, located just west of the Hudson River, penetrated 1900 feet of Marcellus strata. Siltstone beds appear in the gray, brown and black shales of the Marcellus in east-central New York and an easterly coarsening is indicated by sandstone beds that occur in the Marcellus Formation is the Catskill Mountains region.

The top of the Marcellus Formation has been recognized by gamma-ray logs in Pennsylvania. In addition, the Little Gap fossil zone in the Lehighton and Palmerton quadrangles (Epstein and others, 1974) may be equivalent to the Stafford-Mottville interval as may an unnamed limestone 500 feet above the base of the Marcellus at Dalmatia (Faill and others, 1978).

The Marcellus Formation is divided in this report into upper and lower parts, and an isopach map has been drawn for each division. The lower part of the Marcellus includes the Union Springs Shale and overlying Cherry Valley Limestone (Figure 3). In eastern New York these are replaced by the Bakoven Shale and overlying Stony Hollow Sandstone. Correlation into eastern Pennsylvania suggests that the equivalents of these New York units are the Shamokin Shale and overlying Turkey Ridge Member (Faill and others, 1978). The "Selinsgrove upper limestone" of White (1883), located approximately 300 feet above the base of the Marcellus at Selinsgrove Junction (Faill and others, 1978), appears to be equivalent to the Cherry Valley Limestone of New York. In west-central Pennsylvania the equivalents are an unnamed shale at the base of the Hamilton Group and the overlying Purcell Limestone. Rickard (1984) regarded that the Bell Shale and Rockport Quarry Limestone of Ontario and Michigan as the respective equivalents of the Union Springs Shale and Cherry Valley Limestone.

This lower division of the Marcellus Formation is more than 500 feet thick along the eastern outcrop belt in New York (Plate 32). It thins uniformly west of the 200 foot isopach to less than 10 feet in western New York and western Pennsylvania where it remains unidentified in large areas. This division thickens to 70 feet further west in Ontario and is assigned to the Bell Shale and Rockport Quarry Limestone.

The upper part of the Marcellus Formation of New York includes the Oatka Creek Shale and its equivalents to the east (Figure 3). This division includes shales, siltstones and sandstones of the Dalmatia Member in Pennsylvania (Faill and others, 1978). The upper part of the Marcellus increases rapidly in thickness from 800 feet in Greene and Ulster Counties, New York, to 1400 feet at the site of the Shaller well in eastern Greene County. Throughout most of western New York and western Pennsylvania it is less than 50 feet thick. It thickens to 150 feet farther west in Ontario and northeastern Ohio where this division is known as the Arkona Shale (Rickard, 1984).

Skaneateles Formation

The name "Skaneateles" is applied to those rocks of the Hamilton Group between the base of the Stafford-Mottville Limestones and the base of the Centerfield Limestone of the overlying Ludlowville Formation (Figure 3). The Skaneateles Formation is lithostratigraphic unit defined by correlatable marker beds. In eastern New York, these key beds are not presently recognized along the outcrop, and the unit becomes part of the Panther Mountain Formation. However, in the subsurface, the Skaneateles can be traced into Green and Ulster Counties on the basis of the definitive, persistent gamma-ray patterns of the Mottville Sandstone and Centerfield Limestone. The Fisher Ridge Member of the Mahantango Formation of east-central Pennsylvania appears to be equivalent to the Skaneateles of New York (Figure 3).

The calcareous, dark gray, brown or black shale of the Skaneateles in western and central New York grades eastward into interbedded gray shale, siltstone and minor black shale. Sandstone beds appear farther east beneath Chenango, Otsego, Delaware, and Sullivan Counties. Non-marine rocks occur in eastern Greene and Ulster Counties.

The Skaneateles Formation is the lowest Hamilton unit in which whose thickness trends differ markedly from those of the entire Hamilton Group. This is particularly noticeable in the west-east trend of the 400, 500 and 600 foot contours (Plate 35). Maximum thickness of the Skaneateles is over 700 feet along the eastern outcrop belt. The formation thins westwardly to less than 10 feet in northwestern Pennsylvania where the unit consists largely of the Stafford Limestone.

Ludlowville Formation

The Ludlowville is another formation defined by key beds, and includes strata bracketed by the base of the Centerfield Limestone and the base of the Tichenor Limestone (Figure 3). Where the Centerfield is no longer recognized along the outcrop in eastern New York, Ludlowville strata are included in the Panther Mountain Formation. However, as the Skaneateles Formation, the Ludlowville can be traced in the subsurface into eastern New York by the definitive gamma-ray patterns of its boundary units, the Centerfield and Tichenor Limestones. The Ludlowville includes the Montebello Sandstone in east-central Pennsylvania and the Gander Run through Clearville Members of the Mahantango Formation in west-central Pennsylvania. A calcareous shale interval 590 feet below the top of the Mahantango reported in north-central Pennsylvania (Faill, 1974) appears to be equivalent to the Centerfield Limestone. Rickard (1984) concluded that the Ludlowville Formation in the Lake Erie region correlated with the Prout Limestone (Ohio) and the Hungry Hollow Formation and overlying Widder Shale and Ipperwash Limestone of Ontario.

Sandstone composes a significant portion of the Ludlowville Formation in both New York (Cortland and Broome Counties and eastward) and Pennsylvania (Snyder and Perry Counties). The Ludlowville consists of non-marine shale and sandstone beneath the Catskill Mountains in eastern New York whereas the Ludlowville of central and western New York is composed of calcareous, dark gray, brown or black shale.

The isopach patterns of the Ludlowville Formation (Plate 36) are distinct from those of the rest of the Hamilton Group (Plate 38). Two depocenters existed near eastern and southern sand sources in Schoharie County, New York, and in central Pennsylvania where the formation is 400 and 800 feet thick, respectively. The formation is over 300 feet thick beneath McKean, Tioga, and Bradford Counties in north-central Pennsylvania but less than 200 feet thick in Lycoming County, Pennsylvania, west-central New York. Plate 36 illustrates the eastern and western limits of the Centerfield Limestone. Equivalent strata in eastern New York and Pennsylvania are calcareous sandstones. Units equivalent to the Centerfield do not exist in the lowermost Ludlowville Formation in Chautauqua County, New York, and western Pennsylvania.

Moscow Formation

The highest unit of the Hamilton Group is the Moscow Formation. The lower boundary of the formation is the base of the Tichenor Limestone in western and west-central New York and the base of the Portland Point Limestone in east-central and eastern New York (Figure 3). Unlike the older Ludlowville and Skaneateles Formations, the Moscow has a persistent unit, the Portland Point Limestone, that can be recognized along the outcrop belt eastward into Greene County (Cooper, 1934; McCave, 1973). In Pennsylvania the equivalent of the Moscow appears to be the Sherman Ridge Member of the Mahantango Formation (Faill and others, 1978). The Tichenor-Portland Point Limestones in the lowest part of the Moscow Formation are represented in central Pennsylvania by the Pokejoy Member (Hasson and Dennison, 1974) and in north-central Pennsylvania by an unnamed, fossiliferous, calcareous interval 425 feet below the top of the Mahantango Formation (Faill, 1979).

The Moscow Formation consists of calcareous, dark gray, brown or black shales throughout western and central New York and Pennsylvania. These rocks are replaced by gray shales and siltstones in Otsego and Delaware Counties, New York. Sandstone beds appear in Schoharie, eastern Delaware, and Sullivan Counties. The Moscow is composed of non-marine sandstones, and red and green shales in the Catskill Mountains. A cherty limestone identified as the "Tichenor-Menteth" is the lateral equivalent of the Moscow Formation in northwestern Pennsylvania and Chatauqua County, New York (see Baird, 1979). In this region, the Deep Run, Kashong, and Windom Members of the Moscow Formation have been replaced by carbonates as well as partially removed by erosion; this results in amalgamation of the Tichenor and Menteth Limestones into a single condensed limestone.

The isopach pattern of the Moscow Formation (Plate 37) is quite different from that of the entire Hamilton Group (Plate 38). The most significant features shown by the isopachs of this region include 1) the large area lacking any Moscow strata in Cambria, Indiana, and Clearfield and Jefferson Counties, Pennsylvania, 2) the thickened Moscow beneath Warren County, Pennsylvania, 3) an area of thinner Moscow in central New York that extends southward into Tioga County, Pennsylvania, and 4) at least three depocenters of deposition along the eastern outcrop belt where the Moscow attains a thickness of 400 or 500 feet. The short dashed lines of Plate 37 show the distribution of the basal carbonates or calcareous shales of the Moscow Formation. The Tichenor Limestone is cherty in western New York, a feature that helps distinguish it from the Tully Limestone that appears in the southwestern corner of the State and in northwestern Pennsylvania. The Tichenor is replaced eastward by the Portland Point Limestone and this, in turn, is replaced by calcareous shale also referred to the Portland Point Member (Baird, 1979). In eastern New York this unit grades into the Cooksburg Sandstone (McCave, 1968, 1973).

Tully Limestone

Heckel (1973) provided a comprehensive study of the Tully Limestone of New York which followed a similar report on the Tully of Pennsylvania (Heckel,

1969). Heckel subdivided the Tully into Lower and Upper Members that are separated by an unconformity. He recognized and named seven beds in the Lower Member and six beds in the Upper Member. The Lower Member consists of calcarenite that grades eastward into calcareous sandstone and siltstone. It is truncated in east-central New York by the Upper Member which consists of buff-weathering calcilutite. In eastern New York, fossiliferous shales and sandstones of the Gilboa Formation are equivalent to the Tully Limestone (Johnson and Friedman, 1969; Johnson, 1972). The Gilboa is replaced along the Catskill Front by non-marine shales and sandstones included in the upper part of the Manorkill Formation.

This report uses Heckel's (1973) definition and subdivision of the Tully Limestone rather than Cooper and Williams' (1935). Following Heckel, the interbedded limestone and black shale sequence that overlies the massive limestone unit is included in the Tully Limestone rather than the overlying Geneseo Formation. Heckel applied the name "Fillmore Glen" to these higher strata. This definition of the Tully is supported by subsurface data that reveal a lateral replacement of the Fillmore Glen shales and limestones by limestone to the south and east.

Heckel's subdivision of the Tully into Lower and Upper Members has not been recognized in gamma-ray logs of the subsurface Tully. However, it is possible to observe the westward thinning and replacement of the West Brook shale bed by the Bellona Coral bed (see Plates 3-8, 17-21). In these cross sections the seven beds of the lower Tully and the basal bed of the upper Tully are undifferentiated. An unconformity is indicated between the Taughannock Falls and Moravia beds where the West Falls shale is absent.

Plate 39 is an isopach map of the Tully Limestone and its eastern equivalent, the Gilboa Formation. This map differs from an earlier one (Jones and Cate, 1957, Figure 2) by excluding beds that belong to the underlying Moscow Formation from the Tully Limestone. Although the Tully Limestone is absent throughout much of western New York, it reappears beneath central Chautauqua County and extends into northwestern Pennsylvania. Wright (1973) misinterpreted this relationship in her study of the subsurface Tully because she confused the Tully and Tichenor limestones in Chautauqua County. In parts of Wyoming and Livingston Counties, Van Tyne et al. (1980) mistakenly identified certain limestone beds in the Windom Member of the Moscow Formation (upper Hamilton) as the Tully Limestone. Gamma-ray logs of wells in these areas do not have the distinctive patterns of the Tully Limestone.

In south-central Pennsylvania, strata equivalent to the Tully Limestone are quite thin and have been included in the Hamilton Group by Faill and others (1978) and Berg and others (1983). However, the fauna of these beds clearly indicates their relationship to the Tully Limestone. Gamma-ray logs of the subsurface Tully agree with this.

The Tully Limestone across New York State varies from 0 to over 100 feet in thickness (Plate 39). Its equivalent in eastern New York, the Gilboa Formation, is about 300 feet thick along the Catskill Front. The Gilboa is dominated by sandstones with minor shales and contrasts with the shale dominated sequences of intervals of the underlying Moscow Formation and overlying Geneseo shale. This feature assists in its recognition of the Gilboa Formation in wells in eastern New York and eastern Pennsylvania. The Tully Limestone is absent in an area that includes eastern Chautauqua and western Cattaragus Counties, New York, and eastern Warren and western McKean Counties Pennsylvania. It is probable that post-Tully and pre-Geneseo erosion, probably subaqueous, removed it along an uplifted area that extended northward into Erie County, New York. The absence of upper Hamilton Group strata in the same area is evidence of this earlier uplift and erosion. The maximum thickness of the Tully Formation along the outcrop belt occurs in southern Lycoming County, Pennsylvania (Heckel, 1969; Faill and others, 1977a, 1977b; Faill, 1979). The subsurface data generally agree with this observation and indicate that the area of thickest Tully extends westward into Clearfield County, Pennsylvania. Smaller areas of abnormally thick Tully occur in northwestern Pennsylvania, in southwestern Bradford County, Pennsylvania, and beneath Allegany County, New York.

Some stratigraphers, especially those of the Pennsylvania Geological Survey, have maintained that the Tully Limestone is present in Crawford and Erie Counties, northwestern Pennsylvania. Others thought that the "drillers Tully" of this area was not the Tully Limestone (Oliver and others, 1967, 1969, Heckel, 1969, Wright, 1973, and Wallace and others, 1977). Careful study of the gamma-ray logs of many wells in southwestern New York and northwestern Pennsylvania not only confirms the presence of the Tully in northwestern Pennsylvania but also indicates that the "drillers Tully" includes the Tichenor-Menteth Limestone as well as highly calcareous beds equivalent to the Moscow Formation. The relatively pure limestone of the Tully is an interval of very low radioactivity whereas the more argillaceous calcareous beds of the Windom and the Tichenor-Menteth limestone have somewhat higher levels of radioactivity (see Plates 9, 13). Gamma-ray logs through the top of the Middle Devonian in this area nicely show this contrast.

ACKNOWLEDGMENTS

Data on well location and gamma ray characteristics of wells (Appendices A and B) are reposited in the oil and gas geophysical records of the New York State Geological Survey. Depth measurements for these wells are recorded in English units and, for this reason, measurements are in feet in this report. The appendices are available as ASCII text on IBM PC 5¼ inch disk, @ $15. Price subject to change. C.E. Brett, E. Landing, and K.B. Miller reviewed the manuscript. E. Landing revised the text for publication.

REFERENCES

Abel, K.D., and Heyman, L. 1981. The Oriskany Sandstone in the subsurface of Pennsylvania. Pa. Geol. Surv., Min. Res. Rept. 81.

Anderson, E.J. 1967. Paleoenvironments of the Coeymans Formation (Lower Devonian) of New York State. Ph.D. diss., Brown University, Providence, R.I.

_______. 1971. Interpretation of calcarenite paleoenvironments. Soc. Econ. Paleontol. Mineral., Eastern Section, Field Conference Guidebook.

_______. 1972. Sedimentary structure assemblages in transgressive and regressive calcarenites. 24th Int. Geol. Cong., Montreal, Sect. 6, p. 369-378.

Arif, A. 1973. Lithofacies and Paleoenvironments of the Becraft Limestone (Lower Devonian) of New York State. M.Sc. thesis, Temple University.

Baird, G.C. 1979. Sedimentary relationships of Portland Point and associated Middle Devonian rocks in central and western New York. New York State Mus. Bull. 433.

_______. 1981. Submarine erosion of a gentle paleoslope: a study of two discontinuities in the New York Devonian. Lethaia 14:105-122.

_______**, and Brett, C.E.** 1981. Submarine discontinuities and sedimentary condensation in the upper Hamilton Group (Middle Devonian): Examination of marine shelf and paleoslope deposits in the Cayuga Valley. *In* New York State Geol. Assn. Guidebook, 53rd Ann. Mtg., SUNY at Binghamton, p. 115-145.

_______**, and** _______. 1983. Regional variation and paleontology of two coral beds in the Middle Devonian Hamilton Group of western New York. J. Paleont. 57:417-446.

_______**, and** _______. 1986. Submarine erosion on the dysaerobic seafloor: Middle Devonian corrasional disconformities in the Cayuga Valley region. *In* New York State Geol. Assn. Guidebook, 58th Ann. Mtg., Cornell University, p. 23-80.

Belak, R. 1980. The Cobleskill and Akron Members of the Rondout Formation: Late Silurian carbonate shelf sedimentation in the Appalachian Basin, New York. J. Sediment. Petrol. 50:1187-1204.

Berg, T.M., McInerney, M.K., Way, J.H., and MacLachlan, D.B. 1983. Stratigraphic correlation chart of Pennsylvania. Pa. Geol. Surv. Gen. Geol. Rept. 75.

Boucot, A.J. 1959. Brachiopods of the Lower Devonian rocks at Highland Mills, N.Y. J. Paleont. 33:727-769.

_______**, Gauri, K.L., and Southard, J.** 1970. Silurian and Lower Devonian brachiopods, structure and stratigraphy of the Green Pond outlier in southeastern New York. Palaeontographica 135:1-59.

Brett, C.E. 1974. Biostratigraphy and paleoecology of the Windom Shale Member (Moscow Formation) in Erie County, New York. *In* New York State Geol. Assn. Guidebook, 46th Ann. Mtg., SUNY College at Fredonia, p. G1-15.

_______, ed. 1986. Dynamic stratigraphy and depositional environments of the Hamilton Group (Middle Devonian) in New York State, *Part I*. New York State Mus. Bull. 457.

_______**, and Baird, G.C.** 1982. Upper Moscow-Genesee stratigraphic relationships in western New York: evidence for regional erosive beveling in the Late Devonian. *In* New York State Geol. Assn. Guidebook, 54th Ann. Mtg., SUNY at Buffalo, p. 19-63.

_______**, and** _______. 1985. Carbonate-shelf cycles in the Middle Devonian of New York: an evaluation of models for the origin of limestones in terrigenous shelf sequences. Geology 13:324-327.

———, ———, **and Miller, K.B.** 1986. Sedimentary cycles and lateral facies gradients across a Middle Devonian shelf-to-basin ramp, Ludlowville Formation, Cayuga Basin. *In* New York State Geol. Assn. Guidebook, 58th Ann. Mtg., Cornell University.

Cate, A.S. 1961. Subsurface structure of plateau region of northcentral and western Pennsylvania on top of the Oriskany Formation. Pa. Geol. Surv. Map 9.

Ciurca, S.J. 1982. Eurypterids, stratigraphy, Late Silurian- Early Devonian of western New York State and Ontario, Canada. *In* New York State Geol. Assn. Guidebook, 54th Ann. Mtg., SUNY at Buffalo, p. 99-120.

Cooper, G.A. 1930. Stratigraphy of the Hamilton Group of New York, Parts I and II. Amer. J. Sci. 19:116-134, 214-236.

———. 1933. Stratigraphy of the Hamilton Group of eastern New York, *Pt. 1*. Amer. J. Sci. 26:537-551.

———. 1934. Stratigraphy of the Hamilton Group of eastern New York, *Pt. 2*. Amer. J. Sci. 27:1-12.

———. 1941. Facies relationships of the Middle Devonian (Hamilton Group) along the Catskill front. (abs.) Geol. Soc. Amer. Bull. 52:1893.

———, **and Warthin, A.S.** 1941. New Middle Devonian stratigraphic names. J. Wash. Acad. Sci. 31:259-260.

———, **and** ———. 1942. New Devonian (Hamilton) correlations. Geol. Soc. Amer. Bull. 53:873-888.

———, **and Williams, J.S.** 1935. Tully Formation of New York. Geol. Soc. Amer. Bull. 46:781-868.

———, **and others.** 1942. Correlation of the Devonian sedimentary formation of North America. Geol. Soc. Amer. Bull. 53:1729-1794.

Copley, D.L., and Gill, Bradley. 1983. The Akron Dolomite (Bass Islands) oil trend in western New York. Ont. Petrol. Inst., 22nd Ann. Conf., London, Ont.

Diecchio, R.J., Jones, S.E., and Dennison, J.M. 1984. Oriskany Sandstone: regional stratigraphic relationships and production trends. W. Va. Geol. Surv. Map WV17.

Dunn, J.R., and Rickard, L.V. 1961. Silurian and Devonian rocks of the central Hudson Valley. New York State Geol. Assn. Guidebook, 33rd Ann. Mtg., Rensselaer Polytechnic Institute, p. C1-28.

Ebright, J.R., Fettke, C.R., and Ingham, A.I. 1949. East Fork-Wharton gas field, Potter County, Pennsylvania. Pa. Geol. Surv., 4th Ser, Bull. M30.

Epstein, A.G., Epstein, J.B., Spink, W.J., and Jennings, D.S. 1967. Upper Silurian and Lower Devonian stratigraphy of northeastern Pennsylvania, New Jersey, and southeasternmost New York. U.S. Geol. Surv. Bull. 1243.

Epstein, C.M. 1970. Lithofacies and paleonenvironments of the Kalkberg Formation (Lower Devonian) of central New York. Ph.D. diss., Brown University, Providence, R.I.

Epstein, J.B. 1984. Onesquethawan stratigraphy (Lower and Middle Devonian) of northeastern Pennsylvania. U.S. Geol. Surv. Prof. Paper 1337.

———, **Sevon, W.D., and Glaeser, J.D.** 1974. Geology and mineral resources of the Lehighton and Palmerton Quad rangles, Carbon and Northampton Counties, Pennsylvania. Pa. Geol. Surv. Atlas 195cd.

Faill, R.T. 1979. Geology and mineral resources of the Montoursville South and Muncy Quadrangles and part of the Hughesville quadrangle, Lycoming, Northumberland, Montour, and Union Counties, Pennsylvania. Pa. Geol. Surv. Atlas 144ab.

———, **Wells, R.B., and Sevon, W.D.** 1977a. Geology and mineral resources of the Linden and Williamsport Quadrangles, Lycoming County, Pennsylvania. Pa. Geol. Surv. Atlas 134ab.

———, ———, **and** ———. 1977b. Geology and mineral resources of the Salladasburg and Cogan Station quadrangles, Lycoming County, Pennsylvania. Pa. Geol. Surv. Atlas 133cd.

———, **Hoskins, D.M., and Wells, R.B.** 1978. Middle Devonian stratigraphy in central Pennsylvania–a revision. Pa. Geol. Surv. Gen. Geol. Rept. 70.

Fernow, L.R. 1961. Paleoecology of the Middle Devonian Hamilton Group in the Cayuga Lake region. Ph.D. diss., Cornell University, Ithaca, N.Y.

Fletcher, F.W., and Woodrow, D.L. 1970. Geology and economic resources of the Pennsylvania portion of the Milford and Port Jervis 15-minute quadrangles. Pa. Geol. Surv. Atlas 223.

Grasso, T.X. 1970. Paleontology, stratigraphy, and paleoecology of the Ludlowville and Moscow Formations (upper Hamilton Group) in central New York. New York State Geol. Assn. Guidebook, 42nd Ann. Mtg., SUNY at Cortland, p. D1-22.

Gray, L.M. 1984. Lithofacies, biofacies, and depositional history of the Centerfield Member (Middle Devonian) of western and central New York State. Ph.D. diss., University of Rochester, Rochester, N.Y.

Harper, J.D. 1969. Stratigraphy, sedimentology and paleoecology of the Rondout Formation (Late Silurian), eastern New York State. Ph.D. diss., Brown University, Providence, R.I.

Hasson, K.O., and Dennison, J.M. 1974. The Pokejoy Member, a new subdivision of the Mahantango Formation (Middle Devonian) in West Virginia, Maryland, and Pennsylvania. Proc. W. Va. Acad. Sci. 46:78-86.

Head, J.W., III. 1969. An integrated model of carbonate depositional basin evolution: Late Cayugan (Upper Silurian) and Helderbergian (Lower Devonian) of the central Appalachians. Ph.D. diss., Brown University, Providence, R.I.

Heckel, P.H. 1969. Devonian Tully Limestone in Pennsylvania and comparison to type Tully Limestone in New York. Pa. Geol. Surv. Inf. Circ. 60.

________. 1973. Nature, origin and significance of the Tully Limestone. Geol. Soc. Amer. Spec. Paper 138.

Heyman, L. 1977. Tully (Middle Devonian) to Queenston (Upper Ordovician) correlations in the subsurface of western Pennsylvania. Pa. Geol. Surv. Min. Res. Rept. 73.

Johnsen, J.H. 1957. The Schoharie Formation: a redefinition. Ph.D. diss., Lehigh University, Bethlehem, PA.

________**, and Southard, J.B.** 1962. The Schoharie Formation in southeastern New York. New York State Geol. Assn. Guidebook, 34th Ann. Mtg., Brooklyn College, p. A7-15.

Johnson, K.G. 1972. Evidence for tidal origin of Late Devonian clastics in eastern New York State, U.S.A. 24th Int. Geol. Cong., Montreal, Sect. 6, p. 285-293.

________**, and Friedman, G.M.** 1969. The Tully clastic correlatives (Upper Devonian) of New York State: a model for recognition of alluvial, dune(?), tidal, nearshore (bar and lagoon), and offshore sedimentary environments in a tectonic delta complex. J. Sediment. Petrol. 39:451-485.

Jones, T.H., and Cate, A.S. 1957. Preliminary report on a regional stratigraphic study of Devonian rocks of Pennsylvania. Pa. Geol. Surv. Special Bull. 8.

Kissling, D.L., and Mosshier, S.O. 1981. Subsurface Onondaga Limestone: stratigraphy, facies and paleogeography. *In* New York State Geol. Assn. Guidebook, 53rd Ann. Mtg., SUNY at Binghamton, p. 279-280.

Kreidler, W.L., Van Tyne, A.M., and Jorgensen, K.M. 1972. Deep wells in New York State. New York State Mus. Bull. 418A.

Laporte, L.F. 1967. Carbonate deposition near mean sea level and resultant facies mosaic: Manlius Formation (Lower Devonian) of New York State. Amer. Assn. Petroleum Geol. Bull. 51:73-101.

________. 1969. Recognition of a transgressive carbonate sequence within an epeiric sea: Helderberg Group (Lower Devonian) of New York State. *In* Depositional environments in carbonate rocks. Friedman, G., ed., Soc. Econ. Paleontol. and Mineral. Spec. Pub. 14, p. 98-119.

Laskowski, E.A. 1956. Sedimentary petrology and petrography of the Esopus, Carlisle Center, and Schoharie Formations (Lower Devonian) in New York State. M.Sc. thesis, Rensselaer Polytechnic Institute, Troy, N.Y.

Lindholm, R.C. 1967. Petrology of the Onondaga Limestone (Middle Devonian) New York. Ph.D diss., Johns Hopkins University, Baltimore, Maryland.

Luther, D.D. 1894. Report on the geology of the Livonia salt shaft. Ann. Rept. New York State Mus. 47:217-324.

McCave, I.N. 1968. Shallow and marginal marine sediments associated with the Catskill complex in the Middle Devonian of New York. *In* Late Paleozoic and Mesozoic continental sedimentation, northeastern North America. de Vries Klein, G., ed., Geol. Soc. Amer. Spec. Paper 106:75-107.

________. 1973. The sedimentology of a transgression: Portland Point and Cooksburg Members (Middle Devonian), New York State. J. Sediment. Petrol. 43:484-504.

Mazullo, S.J. 1973. Deltaic depositional environments in the Hamilton Group (Middle Devonian), southeastern New York State. J. Sediment. Petrol. 43:1061-1071.

Oliver, W.A., Jr. 1951. Middle Devonian coral beds of central New York. Amer. J. Sci. 249:705-728.

———. 1954. Stratigraphy of the Onondaga Limestone (Devonian) in central New York. Geol. Soc. Amer. Bull. 65:621-652.

———. 1956a. Stratigraphy of the Onondaga Limestone in eastern New York. Geol. Soc. Amer. Bull. 67:1441-1474.

———. 1956b. Biostromes and bioherms of the Onondaga Limestone in eastern New York. New York State Mus. Circ. 45.

———. 1960. Coral faunas in the Onondaga Limestone of New York. U.S. Geol. Surv. Prof. Paper 400-B:172-174.

———. 1963. The Onondaga Limestone. *In* Stratigraphy, facies changes and paleoecology of the Lower Devonian Helderberg limestones and Middle Devonian Onondaga limestones. Geol. Soc. Amer. Guidebook, Field trip no. 1, New York, pp. 11-16, 20-21, 26-27, 29-33.

———. 1966. Bois Blanc and Onondaga Formations in western New York and adjacent Ontario. New York State Geol. Assn. Guidebook, 38th Ann. Mtg., SUNY at Buffalo, p. 32-43.

———. 1967. Stratigraphy of the Bois Blanc Formation in New York. U.S. Geol. Surv. Prof. Paper 584-A.

———, deWitt, W., Jr., Dennison, J.M., Hoskins, D.M., and Huddle, J.W. 1967. Devonian of the Appalachian Basin, United States. *In* Oswald, D.H., ed., International Symposium on the Devonian System, *Part I*, p. 1001-1040.

———, ———, ———, ———, **and** ———. 1969. Correlation of Devonian rock units in the Appalachian Basin. U.S. Geol. Surv. Oil and Gas Inv. Chart OC-64.

Ozol, M.A. 1963. Alkali reactivity of cherts and stratigraphy and petrology of cherts and associated limestones of the Onondaga Formation of central and western New York. Ph.D. diss., Rensselaer Polytechnic Institute, Troy, N.Y.

Rickard, L.V. 1952. The Middle Devonian Cherry Valley Limestone of eastern New York. Amer. J. Sci. 250:511-522.

———. 1962. Late Cayugan (Upper Silurian) and Helderbergian (Lower Devonian) stratigraphy in New York. New York State Mus. Bull. 386.

———. 1975. Correlation of the Silurian and Devonian rocks in New York State. New York State Mus. Map and Chart Ser. 24.

———. 1984. Correlation of the subsurface Lower and Middle Devonian of the Lake Erie region. Geol. Soc. Amer. Bull. 95:814-828.

———. 1985. Correlation of the subsurface Lower and Middle Devonian of the Lake Erie region: reply. Geol. Soc. Amer. Bull. 96:1218-1220.

Smith, B. 1935. Geology and mineral resources of the Skaneateles quadrangle. New York State Mus. Bull. 300.

Van Tyne, A.M., Kamakaris, D.G., and Corbo, S. 1980. Isopach Map of Hamilton Group. U.S. Dept. of Energy METC/EGSP Series 121.

Waldman, E. 1979. Time significance of gamma-ray peaks: an example from the Helderberg Group (Lower Devonian), New York. M.Sc. thesis, Rensselaer Polytechnic Institute, Troy, N.Y.

Wallace, L.G., Roen, J.B., and deWitt, W., Jr. 1977. Preliminary stratigraphic cross section showing radioactive zones in the Devonian black shales in the western part of the Appalachian Basin. U.S. Geol. Surv. Oil and Gas Inv. Chart OC-80.

Way, J.H. 1972. Depositional environmental analysis of Middle and Upper Devonian sedimentary rocks Catskill Mountain area, New York. Ph.D. diss., Rensselaer Polytechnic Institute, Troy, N.Y.

White, I.C. 1883. The geology of the Susquehanna River region in the six counties of Wyoming, Lackawanna, Luzerne, Columbia, Montour, and Northumberland. Pa. Geol. Surv. Rept. T3.

Wright, N.A. 1973. Subsurface Tully Limestone, New York and northern Pennsylvania. New York State Mus. Map and Chart Ser. 14.

APPENDIX A

Note: Appendices A and B are available as ASCII text on IBM PC 5 1/4 inch disk, @ $15. Price subject to change.

Identification of wells

Key:
- C cable tool
- R mud rotary
- A air rotary
- G gamma-ray
- S lithic sample

New York State

Well	Permit	Lease	Operator	Township	Type	Log
Allegany County						
1	02967	McElroy	Butera Oil	Centerville	C	S
2	04248	Wolfer	NYS Nat. Gas	Hume	A	GS
3	00345	Connors	Filmore-Belmont	Hume	C	S
4	03956	Cook 2	NYS Nat. Gas	Hume	C	GS
5	14253	Houghton College	Econ. Develop.	Caneadea		G
6	00850	Chase	Parsons	Caneadea	C	S
7	13684	Bentley	Col. Gas Transm.	Granger		G
8	09194	Cartwright	Anderson	Allen	R	GS
9	00086	Sheldon	Brendel, et al	Allen		S
10	03952	Atherton Est.	Parsons	Allen		S
11	04025	Clark	Felmont	New Hudson		GS
12	03986	Botens-Swift	Parsons	New Hudson		S
13	11978	Emerson	Cons. Gas Supl.	New Hudson	A	GS
14	04168	Eberl	Bentley-Gregg	New Hudson	C	GS
15	12718	Clancy	Col. Gas Transm.	Birdsall		G
16	12118	Dennis	Cons. Gas Supl.	Burns		G
17	04388	Shay	Great Lakes Expl.	Almond	C	S
18	13549	NYS Ref. Area 3	Nat. Fuel Gas	Almond		G
19	03922	Fleming	Shaw, et al	Angelica	C	S
20	02312	Duffy	Donahue Assoc.	Angelica	C	GS
21	04925	Lytle	Great Lakes Gas	Angelica	C	GS
22	03264	Herdman	Heiser, et al	Angelica	C	S
23	03995	Hillabush	Bentley, et al	Belfast	C	S
24	03856	Smith	Swan-Bell	Cuba		S
25	04463	Burdick	Joyce Pipeline	Alfred	C	GS
26	04777	Angell	Pennzoil	Ward	C	GS
27	05060	Black Est.	Bradley Prod.	Scio		G
28	00265	Black	Penn. Nat. Gas	Scio	C	S
29	07159	Maybee	Prof. Petrol. Expl.	Scio	C	GS
30	07503	Ramsey 1	Prof. Petrol. Expl.	Scio		G
31	08638	Ramsey 2	Prof. Petrol. Expl.	Amity	C	S
32	11762	O'Brien	Col. Gas Transm.	Scio	R	S
33	07760	Roberts	Prof. Petrol. Expl.	Scio	C	GS
34	07947	Krull	Amity Gas	Scio	C	S

Well	Permit	Lease	Operator	Township	Type	Log
35	07739	Middaugh	Amity Gas	Amity	C	S
36	08031	Trenkle	Flanigan, et al	Wirt	A	S
37	00247	Gilbert	Home Gas	Wirt		G
38	04654	Ames	Bradley Prod.	Wirt	C	GS
39	04849	Shaner	Bradley Prod.	Wirt		G
40	00246	Chadwick	Sawyer	Wirt	C	S
41	08034	Shea	Felmont	Wellsville	A	GS
42	09330	Kler	Felmont	Andover	A	GS
43	00055	Clark Est.	Empire	Andover	C	S
44	07844	Porge	Quaker St. Oil Ref.	Alma	C	GS
45	05510	Clair	Quaker St. Oil Ref.	Alma	C	GS
46	04865	Homestead	Flanigan Bros.	Bolivar	C	GS
47	04833	Beers	Flanigan Bros.	Bolivar		G
48	09896	Quig Hollow	Col. Gas Transm.	Andover	C	GS
49	08664	Crandall Est.	Sylvania	Andover		G
50	00020	Crandall	Empire	Independence	C	S
51	00256	Bassett	Empire	Independence	C	S
52	15054	Lynch	Nat. Fuel Gas	Willing		G
53	00044	Graves	Empire	Independence	C	S
54	00009	Wilson	Empire	Willing	C	S
55	00005	Hilligas	Belmont Quad.	Willing	C	S
56	09131	Shear 2	Cunningham Gas	Willing	C	S
57	00007	Ludden	Cunningham Gas	Willing		G
58	15253	Tait	Nat. Gas Storage	Willing		G
59	00439	Fortner	Cunningham Gas	Willing	C	S
60	00463	Barber	Cabot	Willing	C	S
61	00387	Costello	Empire	Willing	C	S
62	00004	Van Campen	Belmont Quad.	Willing	C	S
63	00389	Dean Est.	Empire	Alma	C	S
64	00235	Gadsby	Allegany Gas	Bolivar	C	S

Broome County

Well	Permit	Lease	Operator	Township	Type	Log
1	06636	Smith	Genegantslet	Triangle	C	S
2	04854	Quarella	Genegantslet	Triangle	C	S
3	08342	Gerst	Joyce Pipeline	Triangle	C	S
4	05087	Richards	Fenix & Scisson	Triangle	R	GS
5	01803	Hotchkiss	Lamphere, et al	Maine	C	S

Cattaraugus County

Well	Permit	Lease	Operator	Township	Type	Log
1	08581	Thomasett	Iroquois	Perrysburg		G
2	05327	Fee	Countryside	Dayton		S
3	11723	Kelview	NYS Nat. Gas	Ashford		G
4	11689	Blesy	NYS Nat. Gas	Ashford	R	S
5	12475	Schichtel	Flint	Ashford	R	GS
6	16525	Flickenstein	Amer. Penn. Energy	East Otto		G
7	11195	Hill 1	Bentley	Otto	A	S
8	11067	Hill 3	Amarex	Otto	R	S
9	05313	Pratt	Iroquois	Otto	C	GS
10	10748	Marek-Studley	Amarex	Otto	A	S

Well	Permit	Lease	Operator	Township	Type	Log
11	05204	Dankert	Iroquois	Otto	C	S
12	08610	Manning	Iroquois	Otto	A	S
13	05375	Hebner	Iroquois	Otto	C	S
14	13537	Folkes	Col. Gas Transm.	Mansfield	R	GS
15	04153	Heron	Humble	New Albion	R	GS
16	11099	Hershberger	Col. Gas Transm.	Conewango	A	GS
17	02673	See	Dusenberry, et al	Conewango	C	S
18	11100	Miller	Col. Gas Transm.	Conewango	R	GS
19	11478	Edmunds	Col. Gas Transm.	Freedom		G
20	11479	Carroll	Col. Gas Transm.	Little Valley		G
21	11480	Waite	Col. Gas Transm.	Napoli		G
22	04238	Hotchkiss	Farmers	Randolph	C	S
23	03919	Olean Airport	Felmont	Ischua	C	S
24	03895	Laubacker	Bentley	Ischua	C	S
25	03906	Thrall	Minter	Ischua	C	S
26	03908	Osgood	Minter	Ischua	C	S
27	03909	Doner	Penn Drilling	Ischua	C	S
28	02963	Smith-Tucker	Bozard Hill	Humphrey	C	S
29	03989	Utter	Swan	Olean	C	S
30	03898	Brown	Felmont	Hinsdale	C	S
31	04574	Crosby	Thropp	Hinsdale	C	GS
32	04832	Smith	Felmont	Allegany		G
33	04660	Smith Est.	Thropp	Allegany	C	S
34	04694	Dutton	Thropp	Hinsdale		G
35	05013	Russell	Flanigan Bros.	Allegany	A	GS
36	04557	Struble	Cordner	Olean		G
37	04529	Lunn, et al	Matthews, et al	Olean	C	S
38	04373	Potter	Flanigan	Allegany	C	S
39	03307	Andrews	Flanigan	Allegany	C	S
40	05106	Ackerly Est.	Comaco Oil	Carrollton	C	GS
41	04675	NYS Conservation	NYS Nat. Gas	Great Valley	C	S
42	03857	Seneca Nations 2	Felmont	Carrollton	C	S
43	03263	Seneca Nations 1	NYS Nat. Gas	Carrollton	C	S
44	03900	Seneca Nations 3	Felmont	Carrollton		S
45	03870	DeWolfe & McDowell	Dellwood	Carrollton	C	S
46	03247	Lockwood 1	Felmont	Red House		S
47	03308	Goodrich	Felmont	Red House	C	S
48	04953	Lockwood 8	Felmont	Red House		G
49	03883	Veach	Felmont	Elko	C	S
50	04088	Seneca Nations 3	Devonian	South Valley	C	GS
51	04713	Wolf	Flanigan Bros.	Allegany		G
52	04554	Goldsboro-Matson	Pennzoil	Carrollton	C	GS
53	03257	Graff	Mid Hudson	Carrollton	C	S
54	03314	Keery 1	Black Rock	Red House		G
55	03885	Keery	Flanigan, et al	Red House	C	S
56	10855	Pfouts 25	Newman & Newman	Carrollton	C	S
57	04114	Walker 2	Kroh, et al	Carrollton	C	S
58	04081	Potter	Benedum	Carrollton		G
59	03987	Torrey	Parsons	Olean	C	S
60	04820	Messer	Comaco	Olean	C	GS
61	03992	Kinley 1	Parson Bros.	Allegany		G

Well	Permit	Lease	Operator	Township	Type	Log
62	04169	Fee 72	Kinley	Allegany	C	S
63	09235	Enterprise	Pennzoil	Allegany	R	GS

Cayuga County

Well	Permit	Lease	Operator	Township	Type	Log
1	04365	Johnson	Frankfort	Sennett	C	S
2	11632	Steimle	Venango	Owasco		G
3	11590	Dooty	Venango	Fleming		G
4	04068	Shoemaker	United Prod.	Springport		G
5	04628	Jellinghaus	Ashland	Springport		G
6	15529	Wells College	Aurelius	Ledyard		G
7	00478	Mahaney	Reserve	Ledyard		S
8	12149	Deptuch	Pominex	Summerhill		G

Chautauqua County

Well	Permit	Lease	Operator	Township	Type	Log
1	12330	Dickenson	Gailbert	Hanover		G
2	11956	Bell	Hopkins	Sheridan		G
3	11512	Reliagas	Flint	Sheridan	G	
4	07694	Hills	Dawley Bros.	Pomfret	C	S
5	11122	Conti Bros. 5	Flint	Arkwright		G
6	09941	Essex	Flint	Pomfret	R	S
7	09939	Lesch	Flint	Pomfret	R	S
8	10375	DeMaggio	Flint	Pomfret	R	S
9	09942	Noble	Flint	Pomfret	R	S
10	09355	Newman	Linn	Pomfret	R	S
11	04986	McGraw	Iroquois	Pomfret	C	S
12	09940	Bennett	Flint	Pomfret	R	GS
13	04948	Wheeler	Iroquois	Pomfret	C	S
14	09943	Gatto	Flint	Portland	R	S
15	09619	Travis	Pyramid	Portland	R	S
16	10536	Cliffstar 4	Flint	Portland	R	S
17	12676	Wilkosz	Flint	Westfield		G
18	02672	Kiester	Lapage	Westfield	C	S
19	12679	LaPorte	Envirogas	Westfield		S
20	08757	Houck 4	Meridian Explo.	Arkwright		G
21	08281	Zahm	Meridian Explo.	Villenova	C	S
22	09193	Fee 2	W.O. & M.L. Smith	Arkwright	C	S
23	05447	Bignell	Trattler & Assoc.	Villenova		GS
24	06745	Smith	W.O. Smith	Arkwright	C	S
25	04154	Shadle	Humble	Cherry Creek	A	GS
26	11897	Cornell	Paragon Resources	Stockton		GS
27	11892	Anderson	Paragon Resources	Stockton		GS
28	11886	Babcock	Paragon Resources	Stockton		GS
29	04161	Breads	R.K. Petrol.	Chautauqua	A	S
30	04190	Tucker	R.K. Petrol	Chautauqua		G
31	09620	Anderson & Jones	Pyramid	Chautauqua	R	S
32	10565	Pettit	Paragon Resources	Chautauqua		G
33	12597	Bertram	Paragon Resources	Chautauqua		G
34	10757	Abbott	Paragon Resources	Chautauqua		G
35	09960	Johnson	Pyramid Petrol.	Ripley	R	S
36	10028	Mooney	Pyramid Petrol.	Ripley		G

Well	Permit	Lease	Operator	Township	Type	Log
37	13544	McDonald	Envirogas	Ripley		G
38	11254	Dieteman	Paragon Resources	Westfield		G
39	11197	Endres	Paragon Resources	Ripley		G
40	11362	Lowes	Paragon Resources	Ripley		G
41	10332	Smith	Paragon Resources	Ripley		G
42	04152	Carnahan	Apache-Hica	Mina	C	GS
43	00878	E. Stetson	Appalach. Devel.	Mina		S
44	15758	Richardson	Flint	Ellery		G
45	04561	Gage	Minard Run	Ellery	A	GS
46	14451	Chautauqua 2	Envirogas	Chautauqua		G
47	14454	Chautauqua Inst. 5	Envirogas	Chautauqua		G
48	11864	Ploss	Resource Explo.	North Harmony		G
49	10931	Kershaw	Col. Gas Transm.	Chautauqua	R	S
50	15182	Greer 2	Envirogas	Chautauqua		G
51	04347	H. Stetsen	Texaco, et al	Mina	C	S
52	11023	Green	Paragon Resources	Mina		G
53	03887	Heckers	Penn Gas	Mina	C	S
54	10915	Carutis	Col. Gas Transm.	Sherman	R	GS
55	11347	Hoyt 3	Resource Explo.	North Harmony		S
56	12805	Cochran	Consol. Gas Supl.	Clymer		G
57	12190	Hitchcock	Consol. Gas Supl.	Clymer		S
58	11705	Legters Bros.	Edgecliff Resources	French Creek		G
59	12703	Gabriel	Consol. Gas Supl.	French Creek		G
60	11667	Peek'n Peek	Edgecliff Resources	French Creek		S
61	12517	Lent	Trahan Petrol.	Cherry Creek		G
62	15521	Bly	Trahan Petrol.	Ellington		G
63	13605	NYS-DEC	Col. Gas Transm.	Gerry		G
64	02964	Barmore 1	Penn Gas	Gerry	C	S
65	02971	Barmore 2	Flanigan	Gerry	C	S
66	01857	Homes	G. Neeley	Gerry		S
67	04437	Harrington	Wolf's Head Ref.	Ellery	A	GS
68	12161	Roraback	Resource Explo.	North Harmony		G
69	12158	Evarts	Resource Explo.	North Harmony		G
70	12127	Cowles 3	Resource Explo.	North Harmony		G
71	11212	Erickson 4	Resource Explo.	North Harmony		S
72	03249	Morse 3	Univ. Delta	Harmony	C	S
73	13696	Raynor, et al	Col. Gas Transm.	Harmony		G
74	10872	Button	Col. Gas Transm.	Harmony	R	GS
75	12582	Chaut. Ref. Area 4	Seis-Ex Geophysical	Harmony		G
76	12586	Chaut. Ref. Area 10	Seis-Ex Geophysical	Harmony		S
77	12930	Caflisch 2	Envirogas	Clymer		G
78	14429	Wassink	Envirogas	Clymer		G
79	05784	Boe	S.T. Musser	Poland	C	S
80	02955	Marsh	Hart, Jeutter & Yeager	Poland	C	S
81	07190	Walker	S.T. Musser	Poland	C	S
82	10873	Warn	Col. Gas Transm.	Carroll	A	GS
83	11038	Sleasman	Col. Gas Transm.	Poland	A	GS
84	14591	Moon Brook C.C.	Envirogas	Ellicott		G
85	00386	Hazzard	Pettigrew	Busti		S
86	00051	Donelson	Pettigrew	Busti	C	S
87	13787	Kidder 2	Appalach. Energy	Busti		G

Well	Permit	Lease	Operator	Township	Type	Log
88	14652	Brown 2	Appalach. Energy	Busti		G
89	14293	Johnson	Envirogas	Busti		G
90	14292	Ewer	Envirogas	Harmony		G
91	14294	Sullivan 2	Envirogas	Harmony		G
92	13916	Sullivan 1	Envirogas	Harmony		G
93	12992	Simmental 3	Bounty	Harmony		G
94	12589	Chaut. Ref. Area 13	Seis-Ex Geophysical	Harmony		G

Chemung County

Well	Permit	Lease	Operator	Township	Type	Log
1	04026	Callahan	Felmont	Catlin		S
2	03933	DeMunn	Trans-Eastern	Catlin		S
3	00509	Treat	Belmont Quad.	Erin	C	S
4	04863	Boor	Rio Oil	Veteran	C	GS
5	04087	Coleman	Felmont	Veteran		GS
6	00599	Turner	Dearth, et al	Veteran	C	S
7	12998	Gregory	Apache	Catlin		G
8	00600	Willis	Peterman, et al	Catlin	C	S
9	00508	Engish	NYS Nat. Gas	Van Etten	C	S
10	00443	Kesselring	NYS Nat. Gas	Van Etten	C	GS
11	04148	Haywood	NYS Nat. Gas	Van Etten		G
12	04350	Pyhtila	NYS Nat. Gas	Van Etten		G
13	03248	Bailey	Felmont	Van Etten	C	S
14	11939	Montgomery	Seis-Ex Geophysical	Erin	R	GS
15	06781	Held	Hammond Assoc.	Erin	C	G
16	10335	Matejka	Shell	Erin		GS
17	02314	Wheaton	Nat. Explo. Devel.	Erin	C	S
18	00512	Struzinsky	Allegany	Erin	C	S
19	00610	Barcus	Updegraff	Elmira	C	S
20	00476	Berthod	Elmira Nat. Gas	Big Flats	C	G
21	03974	May	Hanley & Bird	Southport		G
22	04923	Richards	Rio Oil	Southport	C	GS
23	11931	Grace	Seis-Ex Geophysical	Chemung	R	GS

Chenango County

Well	Permit	Lease	Operator	Township	Type	Log
1	01160	Lobdell	Bradley Prod.	Columbus	C	GS
2	10607	Wahlberg	Amoco Prod.	Guilford	A	GS
3	10609	Hulbert	Amoco Prod.	Oxford		G
4	10608	Small	Amoco Prod.	Coventry	R	GS
5	05344	Decker 5	Genegantslet	Smithville	C	S
6	04475	Decker 1	Genegantslet	Smithville		S
7	11543	Tarbell	Joyce	Smithville	C	S

Cortland County

Well	Permit	Lease	Operator	Township	Type	Log
1	00644	Butler	NYS Nat. Gas	Homer	C	S
2	00433	McCloy	NYS Nat. Gas	Cortlandville	C	S
3	04714	Clough	Gulf	Freetown	R	GS
4	00555	Woodward	Penn-York Nat.	Harford	C	S
5	06778	Propane Plant 3	Atlantic-Richfield	Harford	R	S
6	02308	Overbaugh	Fralich-Suburban	Harford	C	S
7	13251	Storage Terminal	Atlantic-Richfield	Harford		G

Well	Permit	Lease	Operator	Township	Type	Log
Delaware County						
1	04455	Leslie	Gulf	Franklin	A	GS
2	04073	Hirsch	United Prod.	Franklin	R	GS
3	00879	Hazlett	Warner, et al	Franklin	C	S
4	04364	Finch	Gulf	Sidney	A	GS
5	04214	Campbell	Gulf	Hamden	R	GS
6	10227	Weickert	Anschutz	Roxbury		G
7	04379	Lanzilotta	Gulf	Roxbury	R	GS
8	10096	Holdridge	Anschutz	Middletown		GS

Well	Permit	Lease	Operator	Township	Type	Log
Erie County						
1	06668	Bethlehem Steel	Bethlehem Steel	Hamburg		G
2	09938	Bethlehem Steel 4	Flint	Hamburg		G
3	04858	Fee 4	Shoellkopf	Hamburg	C	S
4	05207	Ganey	Shoellkopf	Hamburg	C	S
5	04101	Iroquois 807	Iroquois	Evans		S
6	08862	Bennett Road Farm	Ginell	Evans		G
7	12470	Zittel	Zittel	Eden		G
8	05090	Huff	Iroquois	Evans	C	S
9	11335	Dibello	Marmik	Brant		G
10	11645	Winters	Marmik	Brant	R	S
11	16337	Nordblum	Sinclairville	Brant		G
12	04864	Cornwell	Reserve Gas	Alden		S
13	10381	Fee 1	Flint	Alden		G
14	13913	Dudzic	P & S	Drilling Elma		G
15	12564	Sowinski	Flint	Marilla		G
16	15791	Muntt	Sinclairville	Aurora		G
17	08752	Suozzi 2	Suozzi	Aurora		G
18	03902	Garoc Dairy	Garoc Dairy	Aurora	C	S
19	12452	Hudson	Flint	Wales		G
20	16097	Nicoloff	Subsea Oil	Orchard Park		G
21	16095	Schultz	Energy Oil	Boston		G
22	15760	Parysak	Sinclairville	Boston		G
23	04665	Geiger	Iroquois	North Collins		G
24	03269	Iroquois 796	Iroquois	Concord		S
25	15809	Ballard	Flint	North Collins		G
26	11730	Foss	Flint	Wales	R	GS
27	10836	Colby	Amarex	Holland		G
28	10837	Wroblewski	Amarex	Colden	A	S
29	11510	Potter	Cons. Gas Supl.	Concord		G
30	11509	Whiteman	Cons. Gas Supl.	Concord		G
31	11439	Heary	Cons. Gas Supl.	Concord	R	GS
32	12970	Emerling	Flint	Concord		G
33	11002	Brown	Flint	Sardinia	R	GS
34	04530	LaScala	Flint	Sardinia		G

Well	Permit	Lease	Operator	Township	Type	Log
Genesee County						
1	00652	Edzold	Edzold	Darien	C	S
2	04137	Labenski	Reserve	Darien	C	S

Well	Permit	Lease	Operator	Township	Type	Log
3	13681	Shaffer 6	Reserve	Darien		G
4	05213	Ball	Duchscherer	Pembroke	C	GS
5	12059	Bennett	U.S. Gypsum	Darien	R	G
6	10776	Belt	Flint	Bethany		GS
7	10388	Torrey	Amarex	Bethany		S
8	10387	Cone	Amarex	Stafford		S
9	04906	Niedick	Weaver	Pavillion	C	S

Greene County

Well	Permit	Lease	Operator	Township	Type	Log
1	03904	Gans	United Product.	Windham		GS
2	04750	Borden	Gas, Oil and Land	Catskill	C	S

Herkimer County

Well	Permit	Lease	Operator	Township	Type	Log
1	03993	Skramko	Benedum-Trees	Warren	C	GS

Livingston County

Well	Permit	Lease	Operator	Township	Type	Log
1	04567	Johnson	NYS Nat. Gas	Caledonia	A	GS
2	04218	Walton	NYS Nat. Gas	York		G
3	04069	McDonald	NYS Nat. Gas	York	C	GS
4	04552	McClurg	NYS Nat. Gas	York	A	S
5	04531	Hunn	Ventura	York		G
6	04391	Parnell	NYS Nat. Gas	York	C	GS
7	04495	Smith 3	NYS Nat. Gas	York		G
8	04138	DeLaverne	NYS Nat. Gas	Leicester		G
9	03282	Retsof (Luther)	Retsof Salt Mining	York		S
10		Retsof (Fuller)	International Salt	York		S
11	04089	Yunkers	NYS Nat. Gas	Leicester		G
12	04117	Senft	NYS Nat. Gas	Leicester		G
13	04451	Mulligan	Eason	Avon	C	S
14	04149	Wadsworth	NYS Nat. Gas	Avon		G
15	04213	Wadsworth 2	NYS Nat. Gas	Geneseo		G
16	04217	Christiano	Partee	Leicester	C	S
17	11403	McBride	Col. Gas Transm.	Leicester	A	GS
18	04056	Marshall	Kissinger, et al	Avon	C	S
19	10920	Adair	Cons. Gas Supl.	Geneseo		G
20	04188	Austin	NYS Nat. Gas	Geneseo	C	GS
21	13870	Wadsworth	Bennett	Groveland		G
22	13700	Hilts	Col. Gas Transm.	Mt. Morris		GS
23	04457	Krebecke	Eason	Lima	C	S
24	10922	Beecher	Cons. Gas Supl.	Livonia		G
25	03277	Livonia Salt	Livonia Salt & Mining	Livonia		S
26	04053	Nivers	United Produc.	Livonia	R	GS
27	03305	Hunt	Livingston	Conesus	C	GS
28	03942	Martucio	Kenroe	Conesus	C	GS
29	04630	Kennedy	Blair, et al	Sparta	A	GS
30	11402	Thompson	Col. Gas Transm.	West Sparta	A	GS

Madison County

Well	Permit	Lease	Operator	Township	Type	Log
1	09578	Shepard	Cons. Gas Supl.	Fenner		GS

Well	Permit	Lease	Operator	Township	Type	Log
2	04085	Doroshenko	NYS Nat. Gas	Eaton	C	GS
3	04185	Bikowski	NYS Nat. Gas	Eaton		G
4	04045	Seeley	NYS Nat. Gas	Lebanon	A	S
5	04031	Finen	NYS Nat. Gas	Lebanon		G
6	04002	Parteko	NYS Nat. Gas	Lebanon		G
7	03970	Branagan	NYS Nat. Gas	Lebanon	C	GS
8	06782	Upham	Cons. Gas Supl.	Georgetown		G
9	04499	Cook	NYS Nat. Gas	Madison		G
10	04556	Helms	NYS Nat. Gas	Brookfield		G
11	04033	NYS Consv.	NYS Nat. Gas	Brookfield		G
12	03981	Head	NYS Nat. Gas	Brookfield		GS
13	04510	Albee	NYS Nat. Gas	Madison		G
14	04032	Danisevich	NYS Nat. Gas	Brookfield	C	GS
Oneida County						
1	03955	Northrup	NYS Nat. Gas	Sangerfield		G
2	03928	Keith	NYS Nat. Gas	Sangerfield		GS
Onondaga County						
1	12163	Harrison	Pominex	Marcellus	R	GS
2	11655	Frost	Pominex	Marcellus	R	GS
3	04902	Frost	Reliance	Marcellus	C	GS
4	11946	Fisher	Pominex	Onondaga		G
5	11654	Sears	Pominex	Otisco	R	GS
6		Solvay M21	Solvay Corp.	Tully		G
7	12148	Van Patten	Pominex	Tully	R	GS
Ontario County						
1	04450	Gonsenhauser	Eason	West Bloomfield	C	S
2	12195	Champlin	Nat. Explo.	Bloomfield	R	GS
3	04395	Purdy	Eason	Canandaigua	C	S
4	04402	Neenan	Eason	East Bloomfield	C	S
5	04159	Murphy	Allegany Prod.	East Bloomfield	C	S
6	04057	Birx	Joyce Pipeline	East Bloomfield		G
7	04409	Outhouse	Eason	Canandaigua	C	S
8	04947	Pierce	Duchscherer & Strudwick	Canandaigua	R	GS
9	04035	Gladding	Franchot	Bristol	C	GS
10	04027	Mills	Franchot	Bristol		G
11	03991	Kage	Franchot	Bristol		G
12	03918	Rolsin	Hurtt	Bristol		G
13	03947	Blackmer	Franchot	Richmond		G
14	04005	Allen	Franchot	Richmond		G
15	03988	Ace	Drury	Richmond		G
16	03997	Clement	Franchot	Richmond		G
17	03971	Treble	Franchot	Richmond		G
18	12230	White	Fitzgerald Bros.	Richmond		G
19	12008	Hamilton	Fitzgerald Bros.	Richmond	R	S
20	03866	Hale	Drury	South Bristol	C	S
21	04449	Bill	NYS Nat. Gas	Hopewell		S

Well	Permit	Lease	Operator	Township	Type	Log
22	11428	Halpin	Minerals Mgt.	Seneca	R	GS
23	06395	Frankish	Hoover	Gorham	R	GS
24	04160	Fordon	Williams	Seneca	C	GS
25	09192	Robson-Onderdonk	Winwest Oil	Seneca	C	S
Otsego County						
1	10138	Schwerd	Cons. Gas Supl.	Burlington		GS
2	04050	Elliott	Prensett	Pittsfield	C	GS
3	02612	Matteson	Blue Rock	Morris	C	S
4	10725	Bellows	Amoco	Morris		GS
5	10838	Johnson	Amoco	Unadilla		GS
6	10834	Hoose	Amoco	Laurens	R	GS
7	04055	Lum	NYS Nat. Gas	Worcester	C	GS
8	04429	Russ	NYS Nat. Gas	Maryland		G
9	04245	Baum	NYS Nat. Gas	Maryland	C	GS
10	04547	Burkhard	NYS Nat. Gas	Maryland	C	GS
Schoharie County						
1	10263	Dayton	Cons. Gas Supl.	Jefferson		GS
Schuyler County						
1	00538	Price	Belmont Quad.	Tyrone		G
2	00561	Best 2	Belmont Quad.	Tyrone		G
3	01118	Olezeswski	Wittmer	Tyrone		G
4	03940	Watkins Glen 29	Internat. Salt	Reading		G
5	13796	Pratt	Col. Gas Transm.	Reading		G
6	00534	Webb	Tri-States	Orange	C	S
7	11921	Sugar Hill Rec. 2	Seis-Ex Geophysical	Orange	R	S
8	03254	NYS Lands 2	Felmont & Donovan	Orange	C	S
9	00532	Kellogg	Burdette	Hector		S
10	04187	Morris	Felmont	Hector	C	S
Seneca County						
1	04600	Christensen	NYS Nat. Gas	Fayette		G
2	04215	Garrett	NYS Nat. Gas	Fayette		G
3	04203	Schaffer 2	United Prod.	Fayette		GS
4	04082	Stein	Benedum	Fayette	C	GS
5	04064	DeWall	United Prod.	Fayette		G
6	10960	Warren Bros.	Ashland	Fayette	C	GS
7	04244	Robson	NYS Nat. Gas	Fayette		G
Steuben County						
1	09359	Magin	Eastern States	Dansville	C	S
2	10024	Fuller	Cons. Gas Supl.	Dansville	A	GS
3	11135	Pillen	Cons. Gas Supl.	Dansville		G
4	11754	Blades	Seis-Ex Geophysical	Hornell	R	GS
5	11027	Mann	Cons. Gas .Supl.	Wheeler		G
6	13736	Wagner	Col. Gas Transm.	Wheeler		G

Well	Permit	Lease	Operator	Township	Type	Log
7	00143	House	Liederbach, et al	Howard	C	S
8	00138	Freeborn	Penn York & Empire	Howard	C	S
9	04474	Mulhollen	Fralich, et al	Hornellsville		S
10	00042	Falzarano	Appal. Gas, et al	Hornellsville	C	S
11	00081	Hemphill	New Penn. Devel.	Hornellsville	C	S
12	08045	Marvin & Wilkins	Home Gas	Hartsville	A	S
13	12789	Parker	G.W. Gas & Oil	Howard	C	S
14	11077	Richardson 2	J.V.P. Resources	Howard	R	GS
15	10334	Richardson 1	J.V.P. Resources	Howard	R	GS
16	11126	Morse	Felmont	Hartsville	R	GS
17	09572	Woodworth	Nat. Petrol.	Hartsville	C	S
18	07738	Kurtzworth	Prof. Petrol. Explo.	Hartsville	C	S
19	03864	Belmont Quad.	Felmont & Donovan	Hartsville	C	S
20	10036	Col. Gas H-248	Col. Gas Transm.	Hartsville	R	GS
21	10246	Col. Gas H-255	Col. Gas Transm.	Hartsville	R	GS
22	09541	Mullen	Col. Gas Transm.	Hartsville	A	S
23	13699	NYS Reforestation	Col. Gas Transm.	Wheeler		G
24	11570	Brayer	Cons. Gas Supl.	Pulteney	R	GS
25	05063	Smolos	Great Lakes, et al	Pulteney	C	GS
26	11561	Taylor Wine	Cons. Gas Supl.	Urbana		G
27	00326	Wood	Jackel, et al	Urbana	C	S
28	00569	Bliss	Reserve	Wayne		G
29	11093	Green	Cons. Gas Supl.	Bath		G
30	12402	Kish	Anderson	Bath		S
31	08348	Sylvania	Cabot	Bath		S
32	09582	Mullins	Anderson	Bath	R	GS
33	10107	Wood	Anderson	Canisteo	R	GS
34	01807	Allen	Acme Drilling	Canisteo	C	S
35	08318	Raish	Cabot	Canisteo	A	GS
36	11105	Scott	Cons. Gas Supl.	Canisteo		G
37	11044	Lamphier	Green Hill Assoc.	Canisteo	R	S
38	05377	Bassett	Lenco Drilling	Canisteo	C	S
39	00558	NYS 3	Appal. Devel.	Greenwood	C	S
40	00120	Scribner	Home Gas	Greenwood		G
41	00033	Weber	Home Gas	Greenwood		GS
42	00037	Warriner	Wittmer	Greenwood	C	S
43	00434	Warriner	Appal. Devel.	Greenwood	C	S
44	00026	NYS 2	Home Gas	Greenwood		G
45	00036	Clark 2	Home Gas	Greenwood		G
46	00025	NYS 1	Home Gas	Greenwood	C	GS
47	00053	Rogers	Empire	Greenwood	C	S
48	11608	Button	Marsh Hill Assoc.	Jasper	R	S
49	07500	Waufle	Sylvania	Jasper	A	S
50	12179	Hill	Anderson	Cameron		GS
51	02679	Blair	Anchor Petrol.	Bath	C	S
52	01139	Whitcomb	Donovan, et al	Bath	C	S
53	09535	Moran	Flanigan Bros.	Bath	C	S
54	08027	Sutherland	Flanigan Bros.	Bradford	C	S
55	11781	Moran	Seis-Ex Geophysical	Bath		G
56	00073	Carthley	NYS Nat. Gas	Bradford	C	S
57	10259	Cole	Murphy	Campbell	R	S

Well	Permit	Lease	Operator	Township	Type	Log
58	04573	Scudder	Harding, et al	Campbell	C	GS
59	00064	Pooley	Peterman, et al	Campbell	R	S
60	11462	Lyle	Thurston Assoc.	Thurston	R	GS
61	07510	Naar	Sylvania	Cameron	A	GS
62	00660	Taft	Belmont Quad.	Cameron	C	S
63	00097	Federal Land Bank	Cabot	Cameron	C	S
64	00156	Walwrath	New Penn Devel.	Jasper	C	S
65	00048	House	NYS Nat. Gas	Jasper	C	S
66	00149	Wallace	Sylvania	Jasper	C	S
67	05278	Guild	Wyckoff Devel.	Jasper	A	GS
68	05123	Cornell	Wyckoff Devel.	Jasper	C	GS
69	05549	Van Wie	Wyckoff Devel.	Jasper	A	GS
70	10369	Manning	Reetz Assoc.	Jasper	R	S
71	05555	Dibble	Wyckoff Devel.	Jasper		G
72	05315	Chase	Sylvania	Jasper		G
73	05288	Banks	Sylvania	Jasper		G
74	03897	White	Allegany Prod.	Troupsburg	C	S
75	10525	White	Cabot	Troupsburg		G
76	09138	White	Messer	Troupsburg	C	S
77	00413	McKeon	Sylvania	West Union	C	S
78	00128	McCormick	Rex Nat. Gas	Greenwood	C	S
79	00210	Crane 1	NYS Nat. Gas	Woodhull		G
80	00211	Crane 3	Cabot	Woodhull	C	S
81	00216	Harrington	New Penn Devel.	Woodhull		S
82	03924	Olin	NYS Nat. Gas	Woodhull		GS
83	03894	Hargrave	NYS Nat. Gas	Woodhull		GS
84	00047	Grist	NYS Nat. Gas	Woodhull	C	S
85	13011	Burdick	Col. Gas Transm.	Troupsburg		G
86	08830	Good	Minter, et al	West Union	A	S
87	00290	Hendrickson	Sylvania	Troupsburg	C	S
88	00291	Schall	NYS Nat. Gas	Troupsburg		G
89	01809	Stebbins 2	Appal. Devel.	West Union	C	S
90	03896	NYS 3	Donovan	Hornby	C	S
91	03246	NYS, et al	Felmont	Hornby	C	S
92	02965	Roloson	Cavalier	Hornby	C	S
93	03932	Stevens	Leiderbach	Hornby		S
94	00058	Alger	Corning Glass	Corning		S
95	03250	Thrall	NYS Nat. Gas	Corning		S
96	11942	Scudder	Gailbert Nat. Gas	Hornby	R	S
97	12446	Cass	J.V.P. Explo.	Hornby		S
98	13243	Leach	Donohue, et al	Hornby		G
99	13063	Erwin Pond	Cons. Gas Supl.	Erwin		G
100	00111	Lilley	Empire Keystone	Erwin	C	S
101	04575	Dann	Joyce Pipeline	Erwin	C	S
102	00109	Collins	Cabot	Erwin		S
103	00645	Papa	NYS Nat. Gas	Erwin	C	S
104	00391	Cole	NYS Nat. Gas	Corning	C	S

Sullivan County

1	08578	Shandalee	Texas Eastern	Rockland	R	GS
2	12861	O'Donnell	Gulf	Thompson	A	GS

Well	Permit	Lease	Operator	Township	Type	Log
Tioga County						
1	00257	Stevens	NYS Nat. Gas	Spencer	C	S
2	09557	Klossner	Shell	Tioga	R	GS
3	09848	Robinson	Shell	Barton	A	GS
4	00800	Pompelly	Susquehanna	Owego	C	S
Tompkins County						
1	03938	Internat. Salt 20	Internat. Salt	Lansing		G
2	13173	Cargill Salt	Cargill Salt	Lansing		G
3	05017	Agard	Great Lakes	Ulysses	C	GS
4	04130	Grund	NYS Nat. Gas	Enfield	A	GS
5	10243	Place	Shell	Enfield	A	GS
6	04467	Richardson	NYS Nat. Gas	Newfield	A	GS
7	03973	Shepherd	NYS	Nat. Gas Danby		GS
8	04007	Smiley	NYS	Nat. Gas Danby	A	GS
9	04446	Turk	Bentley & Fralich	Danby	C	S
Ulster County						
1	03199	Herdman	Dome Gas & Oil	Shandaken	C	GS
2	04202	Schaller	Gas & Oil Land Leasing	Saugerties	C	GS
Wyoming County						
1	13994	Stedman	Lenape Res.	Bennington		G
2	04432	Schuelte	Great Lakes Explo.	Bennington	C	S
3	10917	Feuz	Flint	Sheldon	R	S
4	10939	Metz	Flint	Sheldon		GS
5	04464	Cox	Trans. Amer.	Middlebury		G
6	04536	Page	Trans. Amer.	Middlebury		G
7	04436	Wellman	Trans. Amer.	Middlebury		G
8	11312	Weber	Marmick	Attica	R	GS
9	04447	Warren	Trans. Amer.	Middlebury		G
10	07278	Welch Bros.	Meridian Explo.	Middlebury		GS
11	12178	Meeder	Cons. Gas Supl.	Middlebury		S
12	07234	Farrell	Linn	Middlebury	A	S
13	10935	Hofert	Cons. Gas Supl.	Orangeville		G
14	04392	Werner	NYS Nat. Gas	Orangeville		S
15	10936	Wagenblass	Cons. Gas Supl.	Warsaw		G
16	11311	Kopczynski	Marmick	Sheldon	R	GS
17	13082	Kibler	Midnight	Sheldon		S
18	13084	Galligan	Midnight	Java		S
19	13725	Sicliari	Stead, et al	Java		G
20	13083	Keem	Midnight	Java		S
21	14165	Zielinski	Lenape Res.	Java		G
22	13727	Kepp	Nord Montara	Arcade		G
23	13085	Meyer	Midnight	Java		S
24	14534	Murcin 2	Nord Montara	Arcade		G
25	13790	Zielinski	Stead, et al	Arcade		G
26	13278	George	SCG Gas Quest	Arcade		S
27	04132	Wallace	NYS Nat. Gas	Covington		G

Well	Permit	Lease	Operator	Township	Type	Log
28	04008	Cornwell	NYS Nat. Gas	Perry	C	GS
29	04343	Bell 1	Partee	Covington	C	S
30	04349	Bell 2	Partee	Covington	C	S
31	04487	Miller	Ventura	Perry		G
32	04468	Winters	Ventura	Perry	A	GS
33	04538	Leroy Water	Ventura	Covington		G
34	06073	Fisher	Flanigan & Whorton	Warsaw	C	GS
35	13241	Carlson	Tamarak	Wethersfield		G
36	13242	Fisher	Tamarak	Gainesville		G
37	04385	Johns	Great Lakes Explo.	Gainesville	C	S
38	04162	Morton Fee 17	Morton Salt	Gainesville	C	S
39	00940	Cummings	Flint	Gainesville	C	S
40	02294	Morton Fee 13	Morton Salt	Castile	C	S
41	03880	Morton Fee 14	Morton Salt	Castile		G
42	04092	Veith	NYS Nat. Gas	Gainesville	R	GS
43	11151	McCormick Farms	Cons. Gas Supl.	Gainesville		G
44	03865	Lockwood	Farmers Oil	Genesee Falls		S
45	13268	Flint	Tamarak	Pike		G
46	13308	Shattuck	Tamarak	Pike		G
47	13869	Wells	Bennett	Castile		G

Yates County

Well	Permit	Lease	Operator	Township	Type	Log
1	04797	Loree 2	C.G.S. Managmt.	Benton	C	GS
2	04795	Jepson	G.G.S. Managmt.	Benton		G
3	09846	Christensen	Baslen	Benton	C	S
4	04796	Borglum	C.G.S. Managmt.	Benton	C	GS
5	09851	Morton Fee 14	Morton Salt	Milo	R	S
6	12813	Morton Disposal 3	Morton Salt	Milo		G
7	12812	Morton Disposal 2	Morton Salt	Milo		G
8	00425	Stoutenberg	Pentaga	Barrington	C	S
9	00631	Weldy	Interstate	Barrington	C	S

Commonwealth of Pennsylvania

Well	Quadrangle	Lease	Operator	Township	Type	Log
Allegheny County						
1	Greensburg D2	Monroe	Huntley & Huntley	Monroeville		G
Armstrong County						
1	Rural Valley E4	Martin	Peoples Nat. Gas	Wayne	A	GS
2	Elders Ridge D47	Heasley	Peoples Nat. Gas			G
Beaver County						
1	Columbiana I5	Metropolitian	Quaker State			G
2	New Castle G10	Cole	Peoples Nat. Gas			G
3	New Castle G	Tennis Heirs	Duff & Galey	South Beaver	C	S
4	Sewickley D2	Aliquippa Works	Jones & Laughlin	Aliquippa	A	GS

Well	Quadrangle	Lease	Operator	Township	Type	Log
Blair County						
1	Altoona E1	Weldon	Peoples Nat. Gas			G
Bradford County						
1	Troy A	Braisted	Mosherville	Wells	C	S
2	Sayre A	Teeter	Nat. Explo. & Devel.	Ridgebury	C	S
3	Sayre A	Carver	Parson Bros.	Ridgebury		S
4	Troy E	Olsen	Duff	Columbia	C	S
5	Sayre I	Kennedy	Keefer, et al	Ulster	C	S
6	Sayre G	Dickenson	Penn-York	West Burlington	C	S
7	Towanda F	Morris	Stanton Water	Orwell	C	S
8	Towanda F	Bonin	Weaver	Orwell		S
9	Towanda I6	Comiskey	Shell	Herrick	R	GS
10	Towanda I3	Lundy	Fairman, et al	Standing stone	A	GS
11	Towanda E	Strickland	Fenix & Scisson	Wysox	R	GS
12	Towanda G	Agricult. School	Penn-United	Wysox		S
13	Monroeton A	Hettick	Penn-York	Towanda	C	S
14	Powell B	Headlee	Shaw, et al	West Burlington	C	S
15	Powell A1	French	California	Granville		GS
16	Monroeton B	Kissell	Shell	Asylum	R	S
17	LeRaysville H	Brown	Penn-York	Stevens	C	S
18	Monroeton I2	Blemle	Pure Oil	Wilmot	A	GS
Butler County						
1	Mercer F	Hochenberry	Mfg. Light & Heat	Mercer		S
2	Foxburg G	Walker, et al	Walker, et al	Parker		S
3	Zelienople F	Rice	Peoples Nat. Gas	Lancaster		S
Cambria County						
1	Patton B1	Leiden	Peoples Nat. Gas			G
2	Patton B	Clearfield Coal	Felmont	Chest	R	S
3	Patton D2	Burley	Kewanee			G
4	Patton D5	Bakerton R. & G.	Peoples Nat. Gas	West Carroll		G
5	Johnstown G12	Griffith	Peoples Nat. Gas	Blackwood		G
6	Johnstown F	Clearfield Coal	Penn. Nat. Gas, et al	Jackson	A	S
7	Johnstown D	Reade 3	Penn. Nat. Gas	Jackson	R	S
8	Johnstown A	Johnstown R. & G.	Penn. Nat. Gas	Jackson	R	S
Cameron County						
1	Emporium H13	Emporium Water	Felmont	Lumber		G
2	Colegrave I	Hercules Powder	Sylvania, et al	Shippen		S
3	Emporium H	Penn. Tract 103	Felmont	Lumber	C	S
4	Driftwood A111	Emporium C. C.	Felmont, et al	Lumber		GS
5	Benezette C277	Pardee 4	NYS Nat. Gas	Shippen		GS
6	Driftwood C	Logue	Dougherty, et al	Grove	C	S
7	Driftwood D	New Bethlehem	Fairman	Gibson	A	S
8		Penn Tract 1	NYS Nat. Gas	Grove		S
9	Driftwood E	Dodge Lumber	Beasley	Gibson	C	S

Well	Quadrangle	Lease	Operator	Township	Type	Log
10	Driftwood D	Eaton	Sylvania	Gibson	C	S
11	Driftwood G	Penn Tract 21	Mfg. Light & Heat	Gibson	C	S
12	Karthaus B	Penn Tract D	Keta Gas & Oil	Gibson		S
Centre County						
1	Snowshoe D	Penn Tract E	Mfg. Light & Heat	Burnside		S
2	Karthaus F	O'Brien	Lexa Oil	Burnside		S
Clarion County						
1	Oil City I2	Dunn-UNG	Linn & Patrick	Ashland		GS
2	Foxburg B	Shirey	Peoples Nat. Gas	Salem	R	S
3	Foxburg B2	Howard	Peoples Nat. Gas	Salem	R	GS
4	Tionesta H	United Nat. Gas	Peoples Nat. Gas	Ashland	R	S
5	Clarion A	Bryner	Dempseytown Gas	Elk		S
6	Clarion B	Mays	Fairman, et al	Clarion	A	GS
7	Clarion D6	Ausler	Fairman Drl.	Piney		G
8	Clarion I8	Peoples 4489	Peoples Nat. Gas		G	
Clearfield County						
1	Penfield C	New Shalmut 3	Devonian	Huston		S
2	Penfield A	Green Glen	Rockton, et al	Sandy	A	S
3	Du Bois F36	Salada	NYS Nat. Gas	Sandy		G
4	Du Bois I	R. & P. Coal	NYS Nat. Gas	Brady	A	S
5	Penfield I	Penn Tract C	Mfg. Light & Heat	Pine	C	S
6	Penfield I280	Bailey	NYS Nat. Gas	Pike		G
7	Punxsutawney C	Irwin	Deemer			S
8	Karthaus A	Mt. Steel	NYS Nat. Gas	Corington	C	S
9	Clearfield F45	Clearfield Trust	NYS Nat. Gas	Girard	R	GS
10	Houtzdale B4	Haupt Heirs	Mfg. Light & Heat	Boggs		G
11	Phillipsburg A	Game Lds. Tr. 78	Parsons Bros.	Graham		S
12	Houtzdale D6	Chase	Cons. Gas Supl.			G
13	Houtzdale D1	Mid Penn Coal	NYS Nat. Gas	Bigler		GS
14	Barnesboro C	Leamer	NYS Nat. Gas	Burnside	A	GS
Clinton County						
1	Renovo West	Foley	Cabot	Leidy	C	S
2	Renovo West E288	L. W. 401	Cons. Gas Supl.			G
3	Renovo West D	Penn Tract 4	NYS Nat. Gas	Leidy		S
4	Renovo West D	Penn Tract 2	NYS Nat. Gas	East Keating	C	S
5	Renovo West C	Penn Tract 15	NYS Nat. Gas	Leidy	C	S
6	Renovo West E	Jents	Susquehanna	Leidy	C	S
7	Renovo West F	Martin	Sheridan, et al	Leidy	C	S
8	Renovo East G	Penn Tract 12	NYS Nat. Gas	Noyes	C	S
9	Snowshoe C	Penn Tract E	Mfg. Light & Heat	Beech Creek	C	S
10	Waterville G	Chatham Water	Mfg. Light & Heat	Gallager		S
11	Renovo East F	Penn Tract 13	NYS Nat. Gas	Gallager		S
12	Howard A	Penn Tract 102	Phillips	Beech Creek	A	S

Well	Quadrangle	Lease	Operator	Township	Type	Log
Columbia County						
1	Catawissa	Knarr	Farmers	Locust	C	GS
Crawford County						
1	Girard E	Conneaut Farms	Penn Gas	Spring		S
2	Girard D	Sergeant	Penn Gas	Beaver		S
3	Girard G112	Clark	Transamer.			G
4	Girard G	Foro	Felmont	Beaver	C	S
5	Cambridge Spg. G	Gosick	Atlas & Sun Oil	Cussewago	C	S
6	Cambridge Spg. G	Revak	Atlas & Sun Oil	Cussewago	C	S
7	Girard	Clements	James Drl.			G
8	Girard H	Lehman	Hall, et al	Spring	C	S
9	Linesville A16	Hart	Lake Shore Pipe			G
10	Andover C	Carter	McClusky	Conneaut	C	S
11	Linesville B15	Lampman	Lake Shore Pipe			G
12	Union City E6	Morton	Garrett	Bloomfield	A	GS
13	Union City G	Chapin	Potter Devel.	Rockdale		S
14	Cambridge Spg. H10	Hindle	Atlas Explo.	Venango		G
15	Linesville I	Calvin	Sylvania	East Fallowfield	C	S
16	Corry G	Halfast	Albaugh	Rome		S
17	Titusville A3	Matteson	Col. Gas Transm.	Rome		G
18	Townville E3	E.B. Lumber	Col. Gas Transm.	Troy		G
Elk County						
1	Kane F	Wt. 3788 (3737)	United Nat. Gas	Highland	C	S
2	Hallton B	Wt. 3653	Penn Gas	Highland	C	S
3	Mt. Jewett F	Wt. 2612	Pennzoil	Jones		S
4	Crosby G	Stackpole	Linn, et al	Benzinger	R	S
5	Penfield B	Bond	NYS Nat. Gas	Horton		S
6	Benezette E	Dents Run Coal 2	Cons. Gas Supl.	Benezette	A	S
7	Benezette I	Charleroi Mt. Club	Keta Oil & Gas	Benezette	C	S
8	Ridgeway I	Kaull & Hall	Fairman Drl.	Jay	R	S
9	Benezette G	Penn. Tract 84	Shell	Jay		S
10	Clearfield	Penn. Tract 83	Shell	Jay	A	S
Erie County						
1	Fairview G	Block 1, No. 1	NYS Nat. Gas	offshore		S
2	Northeast 24	Block 2, No. 1	NYS Nat. Gas	offshore		GS
3	Erie	Hammermill 2	Dow Indust.			GS
4	Erie H111	Fish	Linn			G
5	Erie I62	Goodwill-Curly	Stephens			G
6	Erie H	Erie Skeet Assn.	Penn Gas	McKean	C	S
7	Girard C38	Barney	Worldwide			G
8	Girard B254	Howard	Cayman			G
9	Girard A319	Osterburg	Cayman			G
10	Girard A325	Boothby 3	Cayman	East Springield		G
11	Girard B45	Leach	World Petrol.	Elk Creek		G
12	Girard D	Roberts	Britton, et al	Conneaut		S

Well	Permit	Lease	Operator	Township	Type	Log
13	Girard D194	White	Ventura			G
14	Girard D34	Borst	Russell-McConnell	Conneaut		GS
15	Cambridge Spg. B20	Weaver	Sunset Int.	McKean		S
16	Cambridge Spg. A19	Mathewson	Sunset Int.	Franklin		G
17	Girard F	Miller	Penn. Gas		A	S
18	Cambridge Spg. D11	Vanco	Redfern & Herd	Washington	C	GS
19	Cambridge Spg. F	Etheridge	Alleg. Indust.	Washington		S
20	Northeast I	Hoag	Tex. Incorp. Domest.	Venango	C	GS
21	Northeast H 30	Denee	Cons. Gas Supl.			G
22	Union City A5	Lenart	Apache-Hica	Waterford		G
23	Union City C4	Tomcho	Apache-Hica			G
24	Corry A27	Follette	Penn. Gas			G
25	Corry A	Sill	Plymouth	Wayne		G
Forest County						
1	Sheffield	Collins	Wasson, et al	Howe		S
2	Tionesta C	Collins & Clinger	Linn	Green	R	S
3	Sheffield G5	Collins & Clinger	Linn & Patrick	Kingsley		G
4	Tidioute I	Levorson	Quaker State	Kingsley	R	S
Indiana County						
1	Punxsutawney E	Bair	Col. Carbon	Banks	A	S
2	Smicksburg I	Wissinger 3	Peoples Nat. Gas	East Mahoning	R	S
3	Smicksburg G33	Beatty	Peoples Nat. Gas	South Mahoning		G
4	Barnesboro E18	Hanwell	Felmont			G
5	Barnesboro D	Lisowitz	Fralich, et al	Pine	C	S
6	Indiana F29	Stewart	Delta Drl.			G
7	Barnesboro G	Bethlehem-Cuba	Col. Carbon, et al	Buffington	R	S
8	Barnesboro D26	Clearfield Coal 3	Felmont	Pine		G
Jefferson County						
1	Marienville	Kittanning St. For.	Linn & Patrick	Heath		G
2	Marienville I	Verstine & Kline 8	Deemer	Eldred		S
3	Hallton I	Humphry Est.	Aladdin	Snyder		S
4	Punxsutawney A62	Zeedick	Consol. Gas Supl.	McCalmont		G
5	Punxsutawney A	North	Minter, et al	Young	A	S
6	Punxsutawney E	Conrad	Col. Carbon	Gaskill		S
Lackawanna County						
1	Pittston F	Richards	Transcontin.	Ransom		GS
Luzerne County						
1	Harvey Lake H	Goodwin	Farmers	Lehman	C	S
2	Shickshinny A	Harrison	Galey	Huntington		S
3	Wilkes-Barre G	Good	United Prod.	Dorrance		S
Lycoming County						
1	Renovo East C15	Penn Tract 1	NYS Nat. Gas			G

Well	Quadrangle	Lease	Operator	Township	Type	Log
2	Antrim H	Penn Tract 75A	Felmont	Brown	A	S
3	Antrim I	Schneider	California	Pine	C	GS
4	Trout Run A	Hughes	Cabot	Cogan House		S
5	Waterville C4	Powell & Day	Felmont	Pine		GS
6	Trout Run A3	Stryker	Felmont			G
7	Warrensville B1	Ging	Delta Drl.	Cascade		GS
8	Warrensville I	Burkheart	Storey	Mill Creek	C	S

McKean County

1	Kinzua A16	Kewanee	Pennzoil	Corydon	R	GS
2	Kinzua	Williams	Hiawatha	Foster		S
3	Kinzua	Mallory	Appal. Devel.	Corydon		S
4	Kinzua	Kinzua Chemical	Appal. Devel.	Corydon		S
5	Kinzua	Mallory	Hiawatha	Corydon		S
6	Kinzua F22	Onofrio	Kaylar Devel.	Corydon	R	GS
7	Kinzua F	West Branch 4	Healy Petrol.	Bradford		S
8	Bradford E	Kendall	Freeman	Bradford	C	S
9	Kinzua	Wt. 3416	South Penn	Lafayette		S
10	Kinzua I	Wt. 3410	Thornton	Lafayette	C	S
11	Kinzua	Central Lumber	Penn-Bradford	Hamilton		S
12	Bradford G	Wilson 18	South Penn	Lafayette		S
13	Bradford E	Minard Run 1	Minard Run	Bradford	A	GS
14	Bradford E11	Minard Run 2	Minard Run	Lafayette	R	GS
15	Smethport	Taylor	New Penn Devel.	Ceres		S
16	Smethport E	Nunn	Allegheny	Liberty		S
17	Smethport	Zerbe	Potter Devel.	Liberty		S
18	Smethport G5	Reeves	Pennzoil	Keating		G
19	Mt. Jewett	Haywood	Atl. Seaboard	Hamlin		S
20	Mt. Jewett B	Manor Real Est.	Linn & Patrick	Sergeant	A	S

Mercer County

1	Shenango A7	Lihan	Fields	West Salem		G
2	Shenango D	Smith	Hickman & Hart	West Salem	C	S
3	Shenango G5	McKnight	Melben	Pymatuning		GS
4	Youngstown C	Cromarti	Sharon Steel	Hickory		S
5	Youngstown C8	Chadderton	Vandenburg	Shenango		G
6	Stoneboro A	Fleck	Peoples Nat. Gas	Sandy Creek	R	S
7	Stoneboro D8	Temple	Peoples Nat. Gas	Lake		G
8	Stoneboro E	Miller	Mercer	Lake	C	S
9	Stoneboro I10	Johnson	United Nat. Gas	Sandy Lake		G

Northumberland County

1	Sunbury F	Hilbish	Matthews, et al	Point	C	S
2	Shamokin D	Fox	Angas	Shamokin		S

Pike County

1	Milford A	Hess	Transcontin.	Shohola		GS
2	Hawley B	Penn Dept. F. & W.	Texaco	Blooming Grove	R	S
3	Ariel I	Haag	Transcontin.	Greene	C	S

Well	Quadrangle	Lease	Operator	Township	Type	Log
Potter County						
1	Coudersport B	Wilson	Cabot	Sharon	C	S
2	Coudersport B	Dickenson	Cabot	Sharon	C	S
3	Genesee A	Hart	Cunningham	Oswayo	C	S
4	Genesee A	Hart 2	Cunningham	Oswayo	C	S
5	Genesee A	Roche	Cunningham	Oswayo	C	S
6	Genesee A	Markey	Cunningham	Oswayo	C	S
7	Gaines A	Schofield	Lycoming Nat. Gas	Harrison		S
8	Gaines A	Ashton	New Penn Devel.	Harrison	C	S
9	Genesee B38	Johnson	NYS Nat. Gas			G
10	Genesee E	Gale	Hammondsport Gas	Allegany	C	S
11	Genesee D	Wells	Slocum	Allegany		S
12	Genesee D	Meister	Lycoming Nat. Gas	Hebron	C	S
13	Coudersport F	Simpson	Appal. Devel.	Hebron	C	S
14	Coudersport	Matteson	Potter Devel.	Hebron		S
15	Emporium A34	Hammermill Paper	Linn	Roulette		G
16	Emporium D	Crawford Est.	Potter Devel.		CR	S
17	Gaines E	Simmons	United Nat. Gas	Hector	C	S
18	Gaines D	Williams	North Branch Devel.	Hector	C	S
19	Genesee E	Lewis	Cabot	Allegany	C	S
20	Conrad B	Lymans & Wynkoop	Sylvania	Ulysses		S
21	Galeton B	Hill	New Penn Devel.	West Branch	C	S
22	Galeton A	Cizek	Pennzoil United	West Branch	A	S
23	Conrad F60	Pa. Tract 181	Linn	Abbott	R	GS
24	Conrad F	Wt. 4689	Allegany, et al	Abbott	C	S
25	Conrad D61	North Penn Gas	Linn	Eulalia		GS
26	Conrad D	North Penn Gas 4763	Fairman	Wharton	A	S
27	Emporium F11	Wt. 4774	Felmont	Sylvania		GS
28	Emporium H	Pa. Dept. F. & W.	Texaco	Portage	A	S
29	Galeton H	Pa. Tract 82	Phillips	Abbott		GS
30	Galeton G	Pa. Tract 81	Phillips	Stewardson	A	S
31	Conrad G	Emporium Lumber	Minter	Wharton	A	S
32	Conrad G	Peters	Cabot	Wharton	C	S
33	Short Run	Pa. Tr. 2, Wt. 4734	Cabot	Wharton		S
34	Renovo West C	Pa. Dpt. F&W Tr. 129	Cons. Gas Supl.	Stewardson	R	GS
Sullivan County						
1	Powell I2	Bruschart	Bluerock Drl.	Forks		GS
2	Eagles Mere F1	Bennett	California	Davidson		GS
Susquehanna County						
1	Le Raysville	Moran	Kewanee			S
2	Le Raysville C	Kelly-Hendrickson	Gas Well Supl.	Choconut	C	GS
3	Susquehanna G	Pease	Shell	Harford		S
Tioga County						
1	Gaines	Davis-Ashton	Anderson	Brookfield		G
2	Gaines C	Simmons	NYS Nat. Gas	Brookfield		S
3	Tioga	Shoemaker	Lycoming Nat.	Lawrence		S

Well	Quadrangle	Lease	Operator	Township	Type	Log
4	Tioga A	Palmer	East Penn Devel.	Farmington	C	S
5	Elkland F	Colegrove	Allegany	Farmington	C	S
6	Elkland C	Hoyt	Lycoming Nat.	Farmington	C	S
7	Elkland B36	Linder	NYS Nat. Gas	Farmington		G
8	Gaines F	Ackley	North Branch Devel.	Clymer	C	S
9	Gaines F	French	East Penn Devel.	Clymer	C	S
10	Gaines E	Rachael	Ryan	Westfield	R	GS
11	Elkland G	Pa. Tract 150	Pennzoil United	Shippen	A	S
12	Tioga I	Davey	East Penn Devel.	Richmond	C	S
13	Antrim B	Noelke	Bug Drl.	Delmar	C	S
14	Antrim B	Campbell	Allegany & Snee	Delmar	C	S
15	Antrim B	Wheeler	Lycoming Nat.	Delmar	C	S
16	Antrim A	Jones	J & J Enterprizes	Shippen		S
17	Galeton C54	Dewey	Amoco Prod.	Gaines	R	GS
18	Galeton C	Dickerson	NYS Nat. Gas	Gaines	C	S
19	Galeton B	Van Tine	Denman	Gaines		S
20	Antrim F	S & S Coal	Amoco	Morris	A	GS
21	Galeton F	Brewster	Bug Drl.	Elk	C	S
Venango County						
1	Franklin B	Burkhardt	Engles	Sugar Creek		S
2	Franklin F	Grant Fee #348	Pennzoil	Sugar Creek	A	GS
3	Hilliards A3	Tasa Coal 4700	Peoples Nat. Gas	Irwin	R	GS
4	Hilliards A6	Mueller	Peoples Nat. Gas	Irwin		G
5	Hilliards A	Tasa Coal 5240	Peoples Nat. Gas	Irwin		S
Warren County						
1	Youngsville C	Anderson	Pettigrew, et al	Sugar Grove		S
2	Corry C32	Christensen	Col. Gas Transm.	Columbus	R	GS
3	Warren A12	Marsh-Childs	Transamerican	Farmington	A	GS
4	Youngsville F	Keester	Penn Gas	Broken Straw		S
5	Youngsville	Spetz	Potter Devel.	Pittsfield		S
6	Corry I	Smith	Mid. East Oil	Eldred	C	S
7	Titusville C	Reeves	Northern Ordnance	Southwest		S
8	Kinzua	Williams	Appal. Devel.	Corydon		S
9	Kinzua A15	Collins	Felmont			G
10	Warren	Erickson	Updegraff	Elk		S
11	Warren F9	Lindbloom	Thornton			G
12	Kinzua	Jamieson	Sylvania	Kinzua		S
13	Warren	Lessor & Launer	Potter Devel.	Kinzua		S
14	Warren	Lessor & Launer	H C Drl.	Kinzua		S
15	Warren I14	Wt. 2693, Tom-Tom	Pennzoil	Meade	A	GS
16	Tidioute B5	Wheeler-Dusenberry	Benedum	Limestone		G
17	Tidioute B	Kapp	Biery & Johnson	Limestone	C	S
18	Tidioute E	Shaw	Biery & Johnson	Limestone		S
19	Tidioute D2	Proper	Breene & Flanigan	Triumph	C	GS
Washington County						
1	Burgettstown F5	Foley	Benedum Trees	Robinson		G

Well	Quadrangle	Lease	Operator	Township	Type	Log
Wayne County						
1	Damascus C	Price	Transcontin.	Damascus		GS
2	Damascus A	Rodolfy	Col. Gas Transm.	Damascus	R	S
Westmoreland County						
1	Greensburg A3	Sloan	Fox-Coen & Sloan			G
2	Freeport H1	Sloan	Peoples Nat. Gas	Washington		G
3	Latrobe D20	Roskovensky	Peoples Nat. Gas	Unity		G
Wyoming County						
1	Meshoppen E	Sheehan	Riebold	Meshoppen	C	S

State of Ohio

Well	Permit	Lease	Operator	Township	Type	Log
Ashtabula County						
1	P-971	Merritt	Bois D'Arc	Kingsville		G
2		Pinney	Kaiser Al. & Chem.	Ashtabula	C	S
3	P-21	Wild	McClure Oil	Saybrook		G
4		Fox	Benedum Trees	Monroe		G
5		Matulis	Felmont	Pierpont		G
6		Mueller	Devonian & Benedum	Dorset		G
7	P-214	Kaderly	Texaco	New Lyme		G
8	P-86	Kellogg	Wehmeyer	Rome		G
9		French & Papp	McConnell	Andover		G
10	P-103	Clark	Northern Nat. Gas	Windsor		G
Columbiana County						
1		Denny	East Ohio	Knox		G
2		Halverstadt	Mfg. Light & Heat	Salem		G
3	P-539	Batzli	Atlas Explo.	Hanover		GS
4	P-718	Lippencott ?	Pominex ?	West		G
5		Burrows	East Ohio	Hanover		G
6		Gilson	Galey, et al	Madison		G
Coshocton County						
1	P-880	Geib	Roberson, et al	Keene		G
Cuyahoga County						
1	P-389	Bradley	Std. Oil of Ohio	City of Cleveland		G
2	P-910	Brookins	Green Gas	Olmsted		G
3	P-865	City of Cleveland	Monarch Energy	Warrensville		G
4	P-907	North Royalton Bd. of Ed.	Appal. Explo.	Royalton		G

Well	Quadrangle	Lease	Operator	Township	Type	Log
Lake County						
1	P-33	Hawgood	Diamond Alkali 204	Painesville		G
2	P-253	Lake Erie College	Petro. Eval. Serv.	Painesville		G
3	P-238	Lubrizol Corp.	Lubrizol Corp.	Willoughby		G
Lorain County						
1	P-1049	Hill	Apex Nat. Gas	Carlisle		G
2	P-910	Toole	Kapp	Grafton		G
Mahoning County						
1	P-121	Miller	Atlas Explo.	Smith		G
Medina County						
1	P-1256	Divoky 2	Wiser	Hinckley		G
2	P-1312	Fuller	Natol	Brunswick		G
3	P-72	Hawley	King	Westfield		G
Portage County						
1	P-85	Wilson	Northern Nat. Gas	Freedom		G
2		Wheaton	East Ohio	Windham		G
3	P-56	Hinman	Hinton	Ravenna		G
4	P-214			Ravenna		G
5	P-482	Schlarb	Collins & Jones	Franklin		G
6	P-40	Heichel	East Ohio	Brimfield		G
7	P-93	Sveda	East Ohio	Charlestown		G
8		Harville	East Ohio	Rootstown	A	S
9	P-411	O'Neil	East Ohio	Paris		G
10		Hauck	Cayman	Paris		G
11	P-77	Franks	Atlas	Atwater		G
Stark County						
1	P-1563			Paris		G
2	P-963	Czekomski	Nat. Gas W. Va.	Paris		G
Summit County						
1	P-103	Wheatley	East Ohio	Northhampton		G
2		Barberton Mine	Longyear Co.			S
3	P-311	Kiminecz	East Ohio	Franklin		G
Trumbull County						
1	P-16	Wengerd	Northern Nat. Gas	Mesopotamia		G
2	P-13	Frandrich	Northern Nat. Gas	Bloomfield		G
3	P-15	Runkle	Northern Nat. Gas	Gustavus		G
4	P-12	Blaney	Dinger	Hartford		G
5	P-128	Miklos	Sharon Energy	Hubbard		G
6	P-236	Liming & Young	Atlas Resources	Weathersfield		G
7	P-242	General Motors	General Motors	Lordstown		G

Well	Quadrangle	Lease	Operator	Township	Type Log
Tuscarawas County					
1	P-794	Borntrager	Nat. Assoc. Pet.	Sugar Creek	G
2	P-676	Sattler	Status Devel.	Sandy	G
Wayne County					
1	P-979	Hohenshil	Wehmeyer	Green	G
2	P-833	Yoder	Heyser	East Union	G
3	P-1169	Sanger	Kubat	Plain	G
4	P-1229	Petersheim	Ohio Fuel Gas	Salt Creek	G

State of West Viginia

Well	Permit	Lease	Operator	Township	Type Log
Hancock County					
1		Minesinger	Humble	Clay	G

Province of Ontario

Well	Permit	Lease	Operator	Township	Type Log
Elgin County					
1	2820	Consumers 13103	Consumers-Amoco	Lake Erie	G
2	3080	Consumers 13124	Consumers	Lake Erie	G
3	2812	Consumers 13076	Consumers-Amoco	Lake Erie	G
4			Baslen	Yarmouth 14-5	GS
5	4417	Consumers 13393	Consumers	Lake Erie 182-X	G
6	4430	Consumers 13406	Consumers	Lake Erie 184-I	G
7	4484	Consumers 13413	Consumers	Lake Erie 155-Q	G
Kent County					
1	4062	Brascan		Howard	G
2			Carr	Raleigh 22-11-5	GS
3	3532	Wise		Howard	G
4	3486	Consumers 13186		Lake Erie 175-D	G
5	4492	Consumers 13157		Lake Erie 176-S	G
6	4543	Consumers 13262		Lake Erie 220-F	G
7	4030	Consumers 13354	Consumers	Lake Erie 240-6	G
8	4025	Consumers 13355	Consumers	Lake Erie	G
9	4029	Consumers 13360	Consumers	Lake Erie 313-J	G
Lambton County					
1	4504	McClure-Union		Dawn	G
2	4524	Ram No. 57		Enniskillen	G
3			Smyth-Barchild	Plympton 18-7	GS

Well	Quadrangle	Lease	Operator	Township	Type Log
Middlesex County					
1			Canadian Delhi	Ekfrid 5-3-SLR	GS
Norfolk County					
1	2718	Consumers 13047	Consumers	Lake Erie	G
2	2777	Consumers 13100	Consumers-Amoco	Lake Erie	G
3	2788	Consumers 13101	Consumers-Amoco	Lake Erie	G
4		Northcal	Northcal Oil	Lake Erie	S

APPENDIX B

Stratigraphic Data from Wells

Depth in feet to top of formation or member.

Key:

Tl	Tully Limestone	Or	Oriskany Sandstone
Gb	Gilboa Formation	Gl	Glenerie Limestone
Mos	Moscow Formation	Held	Helderberg Limestone
Lud	Ludlowville Formation	Rond	Rondout Formation
Skan	Skaneateles Formation	Sal	Salina Group
Mar	Marcellus Formation	TD	Total Depth
ChVl	Cherry Valley Limestone	NR	Not Reached
Sen	Seneca Limestone	NI	Not Identified
MoNd	Moorehouse and Nedrow Limestones	ABS	Absent
Edg	Edgecliff Limestone	F	Faulted
BoBl	Bois Blanc Limestone	NS	Not Sampled
Scho	Schoharie Formation	DF	Derrick Floor
CaCn	Carlisle Center Formation	KB	Kelly Bearing
Esop	Esopus Shale	GL	Ground Level

New York State

Well	Elev	TlGb	Mos	Lud	Skan	Mar	ChVl	Sen	MoNd	Edg	BoBl	Scho	CaCn	Esop	OrGl	Held	Rond	Sal	TD	Well
Allegany County																				
1	1976DF	2436	2477	2515	2677	NI	NI	2889	2913	NI	ABS	ABS	ABS	ABS	ABS	ABS	ABS	3015	5505	1
2	1560GL	2032	2073	2140	2287	2482	NI	2505	2533	2576	ABS	ABS	ABS	ABS	ABS	ABS	2626	2637	7560	2
3	1530GL	1971	2010	2070	2230	2425	NI	2451	NI	NI	ABS	ABS	ABS	ABS	ABS	ABS	2571	2585	5084	3
4	1680GL	2220	2265	2330	2480	2670	NI	2700	2725	2772	ABS	ABS	ABS	ABS	ABS	ABS	2815	2825	7337	4
5	1240GL	1808	1850	1920	2077	2270	2297?	NR												5
6	1501GL	2080	2130	2200	2353	NI	NI	2564	NI	NI	ABS	ABS	ABS	ABS	ABS	ABS	2674	NI	2721	6
7	1820GL	2445	2493	2583	2710	2930	2953	2961	2982	3030	ABS	ABS	ABS	ABS	ABS	ABS	3050	3088	3151	7
8	1925GL	2574	2623	2720	2848	3072	3102	3119	3138	3190	ABS	ABS	ABS	ABS	ABS	ABS	3207	3240	3412	8
9	1620	2340	2404	2486	2623	2821	NI	2856	NI	NI	ABS	ABS	ABS	ABS	2591	ABS	2972	2995	3027	9
10	1880DF	2545	2625	2705	NI	3070	NI	3102	3122	NI	ABS	ABS	ABS	ABS	3192	ABS	3206	NI	3231	10
11	1679DF	2730	2760	2817	3010?	3181	3200	3208	3217	3272	ABS	ABS	ABS	ABS	ABS	ABS	3304	3312	3344	11
12	2080DF	3172	3217	3270	3480	NI	NI	3628	NI	NI	ABS	ABS	ABS	ABS	3754?	ABS	3757	3776	3784	12
13	2130GL	3270	3300	3350	3557	3702	3722	3730	3737	3800	ABS	ABS	ABS	ABS	ABS	ABS	3834	3847	3997	13
14	2010GL	3261	3295	3345	3562	NI	NI	3726	NI	3785	ABS	ABS	ABS	ABS	3823	ABS	3826	3836	3875	14
15	1782GL	2451	2495	2605	2722	2968	2994	3007	3024	3071	ABS	ABS	ABS	ABS	ABS	ABS	3089	3112	3228	15
16		2417	2460	2592	2707	2966	2985	3003	3020	3069	ABS	ABS	ABS	ABS	ABS	ABS	3079	3115		16
17	1880DF	2702	2735	2865	2975	NI	NI	3275	NI	NI	ABS	ABS	ABS	ABS	ABS	ABS	3353	3372	3407	17
18	1839GL	2920	2968	3093	3218	3475	3495	3510	3527	3574	ABS	ABS	ABS	ABS	3585	ABS	3595	3622		18
19	2138GL	3397	3460	3550	NS	3910	3928	3945	3960	NI	ABS	ABS	ABS	ABS	4023	ABS	4036	4043	4046	19

Well	Elev	TlGb	Mos	Lud	Skan	Mar	ChVl	Sen	MoNd	Edg	BoBl	Scho	CaCn	Esop	OrGl	Held	Rond	Sal	TD	Well
20	1504	2793	2857	2940	3087	3300	3317	3333	3349	NI	ABS	ABS	ABS	ABS	3409	ABS	3427		3432	20
21	1610GL	2857	2910	3010	3154	3384	3400	3415	3430	3472	ABS	ABS	ABS	ABS	3488	ABS	3497		3519	21
22	1952GL	3255	3307	3403	3550	NI	NI	3790	NI	NI	ABS	ABS	ABS	ABS	3877	ABS	3893	3900	5475	22
23	1596DF	2871	2908	NS	NS	3345	NI	3361	NI	NI	ABS	ABS	ABS	ABS	3452	ABS	3453	3496	3520	23
24	1820DF	3407	3434	NI	NI	NI	NI	3881	NI	NI	ABS	ABS	ABS	ABS	3959	ABS	3971		3977	24
25	1753GL	3173	3220	3370	3500	3757	3775	3800	3809	3835	ABS	ABS	ABS	ABS	3840?	3850?			3852	25
26	1874GL	3523	3575	3695	3875?	4077	4095	4108	4125	4157	ABS	ABS	ABS	ABS	4172	ABS			4177	26
27	1983GL	3682	3737	3845	4035	4220	4237	4250	4262	4292	ABS	ABS	ABS	ABS	4304	ABS			4308	27
28	1778GL	3465	3520	3632	NI	NI	NI	4032	4050	NI	ABS	ABS	ABS	ABS	4092	ABS			4103	28
29	1485GL	3235	3289	3383	3589	3754	3773	3785	3800	NI	ABS	ABS	ABS	ABS	3845	ABS			3851	29
30		3240	3290	3384	3594	3750	3770	3780	3795	NI	ABS	ABS	ABS	ABS	NI	ABS	NI		3836	30
31	1729GL	3450	3500	NI	NI	NI	NI	4012	NI	NI	ABS	ABS	ABS	ABS	4080	ABS			4091	31
32	2018KB	3790	3845	3940	4162	NI	NI	4353	NI	NI	ABS	ABS	ABS	ABS	4390	ABS	4443		4470	32
33	1614GL	3351	3404	3495	3715	3885	3904	3915	NI	NI	ABS	ABS	ABS	ABS	4000	ABS			4012	33
34	1825GL	3555	3606	3687	NI	NI	NI	4112	NI	NI	ABS	ABS	ABS	ABS	4173	ABS			4177	34
35	2105GL	3792	3856	3950	NI	NI	NI	4355	4365	NI	ABS	ABS	ABS	ABS	4426	ABS	4442	4460	6084	35
36	2092GL	4060	4115	4193	NI	NI	NI	4590	4615	NI	ABS	ABS	ABS	ABS	4664	ABS	4676		4695	36
37		3993																	4169	37
38	2228GL	4108	4150	4210	4463	4587	4606	4620	4646	4673	ABS	ABS	ABS	ABS	4692	ABS	4713		4721	38
39	2038GL	3940	3980	4040	4295	4415	4434	4447	4468	4492	ABS	ABS	ABS	ABS	4510	ABS			4539	39
40	1860GL	3848	3883	3950	4207	4310	4333	4347	NI	NI	ABS	ABS	ABS	ABS	4393	ABS	4397	4406	4850	40
41	2170GL	4300	4348	4484	4638	4855	4875	4890	4903	4930	ABS	ABS	ABS	ABS	4940				4950	41
42	2196GL	4310	4354	4500	4650	4866	4883	4904	4913	4933	ABS	ABS	ABS	ABS	4942				4952	42
43	2138GL	4207	4270	4417	NI	4788	NI	4820	4830	NI	ABS	ABS	ABS	ABS	4874				4879	43
44	2104	4206	4260	4346	4542	4704	4720	4735	4752	4777	ABS	ABS	ABS	ABS	4785	ABS			4789	44
45	1974GL	4077	4127	4213	4413	4568	4587	4600	4617	4640	ABS	ABS	ABS	ABS	4650	ABS			4662	45
46	2185GL	4320	4367	4442	4673	4808	4825	4840	4857	4880	ABS	ABS	ABS	ABS	4900	ABS	4910		4923	46
47	1987GL	4103	4147F																	
		4147	4195	4270	4510	4641	4675	4696	4722	4759	ABS	ABS	ABS	ABS	4774	ABS			4817	47
48	1956GL	3969	4016	4174	4317	4557	4577	4594	4606	NI	ABS	ABS	ABS	ABS	4637				4647	48
49	2335GL	4317	4364	4517	4661	4885	4905	4922	4935	4964	ABS	ABS	ABS	ABS	4970				5052	49
50	2280GL	4248	4305	4450	NI	NI	NI	4870	4880	NI	ABS	ABS	ABS	ABS	4912				4923	50
51	2188GL	4272	4320	NS	NS	NS	NS	4895	NI	NI	ABS	ABS	ABS	ABS	4933				4943	51
52	2090GL	4170	4210	4364F	4572	4800	4818	4842	4854	4875	ABS	ABS	ABS	ABS	4885				4947	52
53	2183GL	4260	4305	NI	NI	NI	NI	5150	5163	NI	ABS	ABS	ABS	ABS	5213				5219	53
54	2221DF	4253	4310	4470	NI	NI	NI	4868	NI	NI	ABS	ABS	ABS	ABS	4904				4920	54
55	2146GL	4190	4240	NI	NI	NI	NI	4775	NI	NI	ABS	ABS	ABS	ABS	4822				4825	55
56	2220GL	4250	4310	NI	NI	NI	NI	4908	NI	NI	ABS	ABS	ABS	ABS	4952				4954	56
57		4310	4357	4507	4660	4877	4895	4918	4930	4950	ABS	ABS	ABS	ABS	4962				5032	57
58	2124GL	4197	4241	4385	4540	4743	4764	4786	4801	4823	ABS	ABS	ABS	ABS	4833				4917	58
59	2272GL	4296	4350	4500	NI	NI	NI	4946	NI	NI	ABS	ABS	ABS	ABS	4998				5003	59
60	1645GL	3772	3823	NI	NI	NI	NI	4378	NI	NI	ABS	ABS	ABS	ABS	4416				4424	60
61	2008GL	4140	4190	NI	4485	NI	NI	4710	NI	NI	ABS	ABS	ABS	ABS	4739				4744	61
62	2091GL	4150	4207	NI	NI	NI	NI	4732	NI	NI	ABS	ABS	ABS	ABS	4750				4754	62
63	1960GL	4000	4045	NI	4400	4535	NI	4587	4605	NI	ABS	ABS	ABS	ABS	4627				4632	63
64	2109GL	4359	4400	4475	4712	4810	NI	4874	NI	NI	4934	ABS	ABS	ABS	ABS	4952			4973	64

Broome County

Well	Elev	TlGb	Mos	Lud	Skan	Mar	ChVl	Sen	MoNd	Edg	BoBl	Scho	CaCn	Esop	OrGl	Held	Rond	Sal	TD	Well
1	1440GL	1414	1462	1695	1975	2470	NI	3095	NI	NI	ABS	ABS	3145	3185	ABS	3217	3480	3572	5907	1
2	1428GL	1453	1500	1720	2010	2500	3050	3121	NI	NI	ABS	ABS	3170	ABS	3200	3205	3470	3555	5835	2
3	1470GL	1487	NS	NS	NS	NS	NS	3173	NI	NI	ABS	ABS	3230	ABS	3267	3272	3510	3590	6420	3
4	996GL	1425	1469	1700	1988	2510	3033	3099	3115	3145	ABS	ABS	3154	ABS	ABS	3175	3440	3525	9640	4
5	968DF	2006	2038	2310	2625	NI	NI	3732	NI	NI	ABS	ABS	3775	ABS	3810	3843			3850	5

Cattaraugus County

Well	Elev	TlGb	Mos	Lud	Skan	Mar	ChVl	Sen	MoNd	Edg	BoBl	Scho	CaCn	Esop	OrGl	Held	Rond	Sal	TD	Well
1	1388GL	ABS	1647	1655	1872	1895	NI	1939	1957	2050	2102	ABS	ABS	ABS	ABS	ABS	2110	2118	5701	1
2	1299GL	ABS	1727	1743	NI	NI	NI	2045	NI	NI	2234	ABS	ABS	ABS	ABS	ABS	2240	2255	3320	2
3	1380GL	ABS	1676	1709	1948	2014	2035	2039	2064	2163	ABS	ABS	ABS	ABS	ABS	ABS	2211	2221	3286	3
4	1340KB	ABS	1445	NI	NI	1793	1822	1825	NI	NI	ABS	ABS	ABS	ABS	ABS	ABS	2010	2040?	3160	4
5	1350GL	ABS	1684	1713	1960	2015	2042	2048	2074	2173	ABS	ABS	ABS	ABS	ABS	ABS	2220	2235	3317	5
6	1378KB	ABS	1784	1817	2061	2107	2135	2140	2162	2260	ABS	ABS	ABS	ABS	ABS	ABS	2310	2325	3410	6
7	1417GL	ABS	1788	1800	NI	NI	NI	2130	NI	NI	ABS	ABS	ABS	ABS	ABS	ABS	2330	NI	3450	7
8	1485GL	ABS	1870	NI	NI	NI	NI	2207	NI	NI	ABS	ABS	ABS	ABS	ABS	ABS	2400	NI	3462	8
9	1569DF	ABS	1915	1942	2187	2220	NI	2253	2275	2385	ABS	ABS	ABS	ABS	ABS	ABS	2437	2445	3460	9
10	1710GL	ABS	2130	2170	NI	NI	NI	2500	NI	NI	ABS	ABS	ABS	ABS	ABS	ABS	2700	NI	3738	10

Well	Elev	TlGb	Mos	Lud	Skan	Mar	ChVl	Sen	MoNd	Edg	BoBl	Scho	CaCn	Esop	OrGl	Held	Rond	Sal	TD	Well
11	1694GL	ABS	2160	2184	NI	2475	NI	2508	NI	NI	ABS	ABS	ABS	ABS	ABS	ABS	2681	NI	3762	11
12	1777GL	ABS	2368	2373	NI	2643	NI	2670	NI	NI	ABS	ABS	ABS	ABS	ABS	ABS	2860	NI	6762	12
13	1723GL	ABS	2259	2270	NI	2568	NI	2603	NI	NI	ABS	ABS	ABS	ABS	ABS	ABS	2770?	2780?	3850	13
14	1725GL	ABS	2477	2503	2765	2807	NI	2835	2845	2934	ABS	ABS	ABS	ABS	ABS	ABS	2984	3000	4093	14
15	1824GL	ABS	2626	2650	2898	2928	NI	2960	2988	3077	3125	ABS	ABS	ABS	ABS	ABS	3130	3140	4240	15
16	1729GL	ABS	2620	2645	2900	2920	NI	2950	2980	3065	3112	ABS	ABS	ABS	ABS	ABS	3120	3130	4302	16
17	1787GL	ABS	2673	2710	2944	2973	NI	3003	NI	NI	3156	ABS	ABS	ABS	ABS	ABS	3165	3185	4318	17
18	1276GL	ABS	2103	2134	2363	2383	NI	2420	2444	2540	2594	ABS	ABS	ABS	ABS	ABS	2598	2615	3835	18
19	1744	2248	2258	2307	2480	2635	NI	2656	2677	2757	ABS	ABS	ABS	ABS	ABS	ABS	2803	2810	4050	19
20	1691	ABS	2695	2725	2982	3032	NI	3056	3068	3152	ABS	ABS	ABS	ABS	ABS	ABS	3200	3209	4472	20
21	1822	ABS	2865	2885	3152	3183	3213	3218	3230	3320	3369	ABS	ABS	ABS	ABS	ABS	3380	3394	4625	21
22	1805DF	ABS	3090	3145	NI	NI	NI	3410	NI	NI	3560	ABS	ABS	ABS	3570	ABS	NS	NS	4802	22
23	2083DF	3388	3418	3470	3700	3797	NI	3820	NI	NI	ABS	ABS	ABS	ABS	3927	ABS	3932	NR	3940	23
24	1973DF	3287	3317	3367	3607	NI	NI	3726	NI	NI	ABS	ABS	ABS	ABS	3818	ABS	NR		3823	24
25	2027GL	3348	3382	3435	NI	NI	NI	3780	NI	NI	ABS	ABS	ABS	ABS	3905	ABS	NR		3915	25
26	2157DF	3511	3539	3600	NI	NI	NI	3957	NI	NI	ABS	ABS	ABS	ABS	ABS	ABS	4070	NR	4100	26
27	2153GL	3486	3520	3575	3808	NI	NI	3919	NI	NI	ABS	ABS	ABS	ABS	ABS	ABS	4030	4041	4052	27
28	1902DF	3222	3232	3282	3525	3615	NI	3642	NI	NI	ABS	ABS	ABS	ABS	ABS	ABS	3751	NR	3772	28
29	1521DF	2935	2955	NS	NS	NS	NI	3395	NI	NI	ABS	ABS	ABS	ABS	ABS	ABS	3488?	3500?	3561	29
30	2010DF	3545	3580	3625	3876	NI	NI	4001	NI	NI	ABS	ABS	ABS	ABS	4097	ABS	4111	NR	4113	30
31	1822GL	3315	3330	3385	3640	3716	3736	3744	3754	3803	ABS	ABS	ABS	ABS	3830	ABS	3833	NR	3853	31
32	1832GL	3398	3405	3460	3723	3791	3809	3814	3825	3867	ABS	ABS	ABS	ABS	3910	ABS	NR		3937	32
33	1892GL	ABS	3428	3465	NI	3818	NI	3834	NI	NI	ABS	ABS	ABS	ABS	3924	ABS	NR		3935	33
34	1850	3410	3420	3474	3740	3812	3830	3835	3848	3894	ABS	ABS	ABS	ABS	3925	ABS	NR		3935	34
35	1630GL	3231	3240	3297	3568	3636	3653	3658	3669	3715	ABS	ABS	ABS	ABS	3744	ABS	3747	3751	3798	35
36	1798GL	3454	3467	3522	3794	3860	3878	3883	3895	3935	ABS	ABS	ABS	ABS					3943	36
37	1750GL	3390	3402	3460	NI	NI	NI	3816	NI	NI	ABS	ABS	ABS	ABS	3915	ABS	NR		3931	37
38	1869DF	3470	3475	3527	NI	3880	NI	3893	NI	NI	ABS	ABS	ABS	ABS	3992	ABS	4000	4025	4043	38
39	1754GL	ABS	3310	3363	NI	3693	NI	3710	NI	NI	ABS	ABS	ABS	ABS	3809	ABS	3812	NR	3828	39
40	1803GL	ABS	3390	3445	3719	3775	NI	3792	3804	3852	ABS	ABS	ABS	ABS	3891	ABS	3892	NR	3910	40
41	2175GL	ABS	3760	3810	NI	4145	NI	4152	NI	NI	ABS	ABS	ABS	ABS	ABS	ABS	4233	4263	4355	41
42	1630GL	ABS	3260	3295	3563	3619	NI	3625	NI	NI	ABS	ABS	ABS	ABS	3726	ABS	3731	NR	3765	42
43	1419DF	ABS	3047	3066	NI	3380	NI	3394	NI	NI	ABS	ABS	ABS	ABS	3491	ABS	NR		3493	43
44	1418DF	ABS	3050	3090	NI	NI	NI	3435	NI	NI	ABS	ABS	ABS	ABS	3544	ABS	3567	NR	3576	44
45	2269DF	ABS	3970	4010	4264	4323	NI	4334	NI	NI	ABS	ABS	ABS	ABS	4430	ABS	NR		4436	45
46	1625DF	ABS	3242	3303	NI	3608	NI	3625	NI	NI	ABS	ABS	ABS	ABS	3728	ABS	NR		3731	46
47	2078DF	ABS	3650	3705	3968	4010	NI	4027	NI	NI	ABS	ABS	ABS	ABS	4133	ABS	4140	NR	4148	47
48	1974GL	ABS	3582	3630	3918	3950	3966	3971	3978	4020	ABS	ABS	ABS	ABS	4068	ABS			4084	48
49	2006DF	ABS	3788	3833	NI	4110?	NI	4126	NI	NI	ABS	ABS	ABS	ABS	ABS	ABS	4218	4240	4273	49
50	1300GL	ABS	2935	2993	3250	3265	NI	3290	3300	3360	ABS	ABS	ABS	ABS	4303	ABS	3408	3425	4675	50
51	1441	3068	3071	3127	3410	3471	3491	3495	3507	3550	ABS	ABS	ABS	ABS	3583	ABS	3587	3620	3702	51
52	1859GL	ABS	3600	3653	3935	3978	3995	3999	4004	4061	ABS	ABS	ABS	ABS	4096	ABS	4112	4155	4179	52
53	2226GL	ABS	4010	4070	4355	NI	NI	4400	NI	NI	ABS	ABS	ABS	ABS	4482	ABS	NR		4484	53
54	1985DF	ABS	3788	3837	4117	4155	4175	4180	4190	4225	ABS	ABS	ABS	ABS	4265	ABS			4279	54
55	1950DF	ABS	3780	3837	NI	NI	NI	4160	NI	NI	ABS	ABS	ABS	ABS	4242	ABS	4275	NR	4278	55
56	1837KB	ABS	3696	3750	4030	4073	4093	4112	NI	NI	ABS	ABS	ABS	ABS	NR	ABS			4157	56
57	1650GL	ABS	NS	3553	NI	3877	NI	3895	NI	NI	ABS	ABS	ABS	ABS	3978	ABS	NR		4003	57
58	2280GL	ABS	4160	4212	4509	4543	4567	4574	4581	4619	ABS	ABS	ABS	ABS	4643	ABS			4654	58
59	1495DF	3385	3410	3462	NS	3805	NI	3833?	NI	NI	ABS	ABS	ABS	ABS	3931	ABS	3957	4030	4058	59
60	2158DF	4257	4267	4324	4640	4683	4703	4711	4719	4756	ABS	ABS	ABS	ABS	4785	ABS	4800	NR	4847	60
61	2381	ABS	4420	4482	4773	4822	NI	4844	4850	4899	ABS	ABS	ABS	ABS	4917	ABS	4933?		4985	61
62	2385GL	ABS	4442	4502	NI	NI	NI	4846	NI	NI	ABS	ABS	ABS	ABS	4914	ABS	4939	NR	4953	62
63	2278GL	ABS	4247	4300	4591	4635	4652	4658	4665	4707	ABS	ABS	ABS	ABS	4728	ABS	4735	4792	11683	63

Cayuga County

1	750DF							35?	NI	NI	ABS	ABS	ABS	ABS	ABS	112	155	210	2000	1
2	880GL					255	330	350	376	395	ABS	ABS	ABS	ABS	409	421	467	520	2476	2
3	750KB					95	145	165	188	210	ABS	ABS	ABS	ABS	227	238	273	325	2212	3
4	669GL						34	55	77	113	ABS	ABS	ABS	ABS	129	140	169	225	1952	4
5	593GL								4	38	ABS	ABS	ABS	ABS	53	65	100	150	1937	5
6	455GL				20	358	402	422	446	475	ABS	ABS	ABS	ABS	482	513	565	614	2626	6
7	824GL		65?	NS	NS	NS	NS	1010?	NI	NI	ABS	ABS	ABS	ABS	1065	1085	1175	1235	6156	7
8	1585GL	698	722	905	1170	1725	1915	1949	1966	1993	ABS	ABS	ABS	ABS	2002	2008	2125	2190	4550	8

Chautauqua County

Well	Elev	TlGb	Mos	Lud	Skan	Mar	ChVl	Sen	MoNd	Edg	BoBl	Scho	CaCn	Esop	OrGl	Held	Rond	Sal	TD	Well
1	690GL	ABS	604	609	807	827	NI	882	900	1003	1065	ABS	ABS	ABS	ABS	ABS	1098	1105	2089	1
2	663	ABS	598	604	800	808	NI	868	885	990	1052	ABS	ABS	ABS	ABS	ABS	1082	1090	2045	2
3	1115GL	ABS	1223	1233	1425	1430	NI	1492	1510	1620	1675	ABS	ABS	ABS	ABS	ABS	1718	1727	2674	3
4	880GL	ABS	1035	1040	NI	1249	NI	1299	NI	NI	1490?	ABS	ABS	ABS	ABS	ABS	1517	NI	2445	4
5	1120GL	ABS	1375	1390	1573	1583	NI	1644	1663	1774	1830	ABS	ABS	ABS	ABS	ABS	1874	1885	2835	5
6	776GL	ABS	925	945	1100	1110	NI	1170	NI	NI	1350?	ABS	ABS	ABS	ABS	ABS	1390	NI	2370	6
7	778GL	ABS	912	NI	NI	NI	NI	1160	NI	NI	1345?	ABS	ABS	ABS	ABS	ABS	1396	NI	5013	7
8	742GL	ABS	850	870	NI	NI	NI	1100	NI	NI	1285?	ABS	ABS	ABS	ABS	ABS	1320	NI	2253	8
9	705GL	ABS	865	NI	1040	1045	NI	1100	NI	NI	1295	ABS	ABS	ABS	ABS	ABS	1360	NI	2228	9
10	620GL	ABS	640	658	NI	NI	NI	920	NI	NI	1120	ABS	ABS	ABS	ABS	ABS	1190	NI	4770	10
11	700GL	ABS	835	871	NI	NI	NI	1080	NI	NI	1272	ABS	ABS	ABS	ABS	ABS	1326	NI	2227	11
12	755GL	ABS	947	958	1121	1131	1195	1203	1222	1335	1387	ABS	ABS	ABS	ABS	ABS	1428	1440	2364	12
13	1158GL	ABS	1351	1380	1544	1548	NI	1613	NI	NI	1795?	ABS	ABS	ABS	ABS	ABS	1840	NI	2780	13
14	775GL	ABS	970	987	1150	1160	NI	1220	NI	NI	1410?	ABS	ABS	ABS	ABS	ABS	1450	NI	2365	14
15	782GL	ABS	1022	NI	1185	1195	NI	1238?	NI	NI	1450?	ABS	ABS	ABS	ABS	ABS	1500	NI	2370	15
16	868GL	ABS	980	1000	1190	1197	NI	1270	NI	NI	1500	ABS	ABS	ABS	ABS	ABS	1555	NI	2570	16
17	698GL	ABS	954	976	1102	1110	NI	1187	1215	1335	1380	ABS	ABS	ABS	ABS	ABS	1428	1442	2429	17
18	675GL	ABS	883	908	1030?	1038	NI	1114?	NI	NI	1312	ABS	ABS	ABS	ABS	ABS	1326	NI	2271	18
19	640KB	910?	925?	940?	NI	1060	NI	1140	NI	NI	1335	ABS	ABS	ABS	ABS	ABS	1380	NI	2396	19
20	1240GL	ABS	1438	1448	1642	1662	NI	1715	1737	1838	1895	ABS	ABS	ABS	ABS	ABS	1918	1929	2897	20
21	1470GL	ABS	1690?	NI	NI	NI	NI	1970?	NI	NI	2165	ABS	ABS	ABS	ABS	ABS	2218	NI	3164	21
22	1477GL	ABS	1725	1743	1950	1970	NI	2021	NI	NI	2210	ABS	ABS	ABS	ABS	ABS	2237	NI	3166	22
23	1605GL	ABS	1927	1940	2155	2165	NI	2225	2250	2357	2415	ABS	ABS	ABS	ABS	ABS	2434	2445	3433	23
24	1420GL	ABS	1780	1798	2000	2007	NI	2050	NI	NI	2239	ABS	ABS	ABS	ABS	ABS	2271		2291	24
25	1606GL	ABS	2043	2055	2267	2283	NI	2337	2357	2467	2520	ABS	ABS	ABS	ABS	ABS	2540	2555	6280	25
26	1414GL	ABS	1925	1954	2120	2130	NI	2193	2213	2330	2382	ABS	ABS	ABS	ABS	ABS	2422	2435	3400	26
27	1377KB	ABS	1882	1917	2070	2078	NI	2147	2173	2285	2335	ABS	ABS	ABS	ABS	ABS	2387	2400	3400	27
28	1531GL	ABS	2090	2120	2260	2266	NI	2334	2358	2470	2520	ABS	ABS	ABS	ABS	ABS	2567	2580	3540	28
29	1353KB	1850	NI	1880	NI	NI	NI	2055	NI	NI	2250?	ABS	ABS	ABS	2300	ABS	2305	NS	3281	29
30	1410	1857	1868	1895	2010	2020	NI	2100	2120	2230	2282	ABS	ABS	ABS	ABS	ABS	2330	2343	3368	30
31	1390GL	NS	NS	NS	NS	NS	NI	2120?	NI	NI	2310?	ABS	ABS	ABS	NS	ABS	NS	NS	3280	31
32	1410GL	1870	1876	1910	2023	2030	NI	2110	2132	2240	2297	ABS	ABS	ABS	ABS	ABS	2350	2360	3354	32
33	1350GL	1830	1837	1870	1980	1987	NI	2068	2088	2197	2255	ABS	ABS	ABS	ABS	ABS	2300	2312	3323	33
34	1385GL	1870	1877	1910	2020	2028	NI	2105	2128	2238	2294	ABS	ABS	ABS	ABS	ABS	2340	2350	3381	34
35	1495GL	2140	NI	NI	NI	NI	NI	2410?	NI	NI	2490	ABS	ABS	ABS	ABS	ABS	2520	2558	3500	35
36	1290GL	1721	1747	1770	1855	1861	1952	1955	1981	2097	2155	ABS	ABS	ABS	ABS	ABS	2193	2222	3185	36
37	675GL	1032	1067	1090	1161	1166	1267	1270	1297	1410	1468	ABS	ABS	ABS	ABS	ABS	1520	1561	2543	37
38	1587GL	2140	2167	2195	2286	2293	2373	2375	2399	2505	2565	ABS	ABS	ABS	ABS	ABS	2610	2630	3623	38
39	1558KB	2140	2177	2205	2283	2290	2180	2385	2406	2510	2570	ABS	ABS	ABS	ABS	ABS	2610	2650	3600	39
40	1405KB	1957	1994	2020	2092	2100	2191	2197	2219	2332	2380	ABS	ABS	ABS	ABS	ABS	2422	2465	3449	40
41	1502GL	2126	2169	2197	2268	2276	2371	2375	2398	2507	2573	ABS	ABS	ABS	ABS	ABS	2618	2653	3703	41
42	1486GL	2118	2163	2187	2250	2260	NI	2355	2377	2493	2545	ABS	ABS	ABS	ABS	ABS	2594	2615	3578	42
43	1455DF	2073	2120	2140	2212	2223	NI	2309	NI	NI	2505	ABS	ABS	ABS	2550	ABS	NR		2562	43
44	1590GL	2280	2291	2333	2467	2477	2543	2545	2567	2680	2728	ABS	ABS	ABS	ABS	ABS	2780	2785	3790	44
45	1524GL	2196	2204	2246	2383	2391	2443	2445	2467	2572	2627	ABS	ABS	ABS	ABS	ABS	2670	2682	6292	45
46	1380	2101	2111	2162	2281	2287	2358	2362	2380	2480	2545	ABS	ABS	ABS	ABS	ABS	2595	2608	3628	46
47	1440	2171	2180	2232	2347	2354	2425	2429	2446	2546	2612	ABS	ABS	ABS	ABS	ABS	2662	2673	3704	47
48	1640GL	2399	2417	2466	2566	2574	2646	2650	2666	2765	2828	ABS	ABS	ABS	ABS	ABS	2871	2903	3900	48
49	1603GL	2290	2320	2370	NI	NI	NI	2540	NI	NI	2720?	ABS	ABS	ABS	ABS	ABS	2750	2780	3763	49
50		2263	2295	2322	2415	2424	NI	2505	2524	2635	2689	ABS	ABS	ABS	ABS	ABS	2730	2753	3770	50
51	1498DF	2165	2208	2235	NI	NI	NI	2398	NI	NI	2600	ABS	ABS	ABS	2656	ABS	2662	2680	3620	51
52	1511GL	2195	2240	2267	2334	2340	2427	2432	2452	2554	2615	ABS	ABS	ABS	ABS	ABS	2662	2697	3730	52
53	1572GL	2358	2420	2445	2505?	2515	NI	2595	NI	NI	2780	ABS	ABS	ABS	2826	ABS	2830	2860	3813	53
54	1760GL	2580	2603	2655	2750	2760	NI	2830	2850	2947	3007	ABS	ABS	ABS	ABS	ABS	3058	3075	4099	54
55	1711KB	2518	NI	NI	NI	2730	NI	2780	NI	NI	2990	ABS	ABS	ABS	ABS	ABS	3048	NI	4070	55
56	1870GL	2844	2884	2945	3018	3026	3100	3102	3123	3214	3274	ABS	ABS	ABS	ABS	ABS	3330	3341	4426	56
57	1755KB	2707	2750	NI	NI	2900	NI	2960	NI	NI	3170	ABS	ABS	ABS	ABS	ABS	3240	NI	4233	57
58	1430GL	2392	2445	2486	2550	2555	NI	2636	2658	2751	2812	ABS	ABS	ABS	ABS	ABS	2874	2929	4099	58
59	1490GL	2454	2507	2544	2611	2615	NI	2696	2717	2805	2862	ABS	ABS	ABS	ABS	ABS	2912	2928	4025	59
60	1410KB	2256	2305	2335	NI	2550	NI	2609	NI	NI	NI	ABS	ABS	ABS	ABS	ABS	2818	NI	3861	60
61	1595GL	ABS	2253	2289	2510	2522	NI	2568	2587	2684	2740	ABS	ABS	ABS	ABS	ABS	2750	2760	3917	61
62	1653GL	ABS	2430	2468	2690	2704	NI	2747	2768	2868	2923	ABS	ABS	ABS	ABS	ABS	2935	2950	4095	62
63	1930GL	ABS	2594	2634	2830	2840	2895	2901	2919	3022	3090	ABS	ABS	ABS	ABS	ABS	3115	3125	4260	63
64	1775DF	ABS	2495	2530	2715	2723	NI	2778	NI	NI	2975	ABS	ABS	ABS	3008	ABS	3010	3023	4089	64
65	1811GL	2514	2518	2570	2768	2775	NI	2832	NI	NI	2995?	ABS	ABS	ABS	3030	ABS	3032	3050	4109	65
66	1340GL	NS	NS	NI	2350	2360	NI	2406	NI	NI	2585	ABS	ABS	ABS	2618	ABS	2620	2635	3751	66

Well	Elev	TlGb	Mos	Lud	Skan	Mar	ChVl	Sen	MoNd	Edg	BoBl	Scho	CaCn	Esop	OrGl	Held	Rond	Sal	TD	Well
67	1760GL	2360	2367	2416	2571	2580	2640	2652	2672	2775	2841	ABS	ABS	ABS	ABS	ABS	2880	2895	4145	67
68	1510GL	2451	2460	2531	2658	2666	2730	2741	2768	2856	2916	ABS	ABS	ABS	ABS	ABS	2948	2959	4048	68
69	1686GL	2680	2687	2761	2888	2898	2958	2967	2988	3070	3135	ABS	ABS	ABS	ABS	ABS	3172	3186	4259	69
70	1395GL	2405	2419	2495	2613	2622	2685	2693	2716	2802	2868	ABS	ABS	ABS	ABS	ABS	2912	2923	4025	70
71	1548KB	2530	2545?	NI	NI	NI	NI	2810	NI	NI	2975	ABS	ABS	ABS	ABS	ABS	2990	NI	3910	71
72	1579GL	2650?	NI	2704?	NS	NS	NI	2941?	NI	NI	3102	ABS	ABS	ABS	ABS	ABS	3148	3165	4258	72
73	1448GL	2465	2483	2561	2670	2678	2740	2745	2768	2853	2924	ABS	ABS	ABS	ABS	ABS	2963	2977	4133	73
74	1552GL	2616	2637	2717	2820	2833	NI	2894	2924	3005	3065	ABS	ABS	ABS	ABS	ABS	3110	3129	4310	74
75	1715GL	2784	2813	2895	2991	2997	NI	3063	3089	3166	3230	ABS	ABS	ABS	ABS	ABS	3264	3281	4429	75
76	1777GL	2835	NI	NI	NI	NI	NI	3160	NI	NI	3370	ABS	ABS	ABS	ABS	ABS	3400	NI	4480	76
77	1460GL	2569	2608	2689	2766	2772	NI	2840	2864	2954	3016	ABS	AS	ABS	ABS	ABS	3047	3072	4245	77
78	1455	2528	2574	2637	2705	2711	NI	2784	2809	2900	2962	ABS	ABS	ABS	ABS	ABS	2994	3017	4221	78
79	1365GL	ABS	2467	2500	NI	NI	NI	NR				ABS	ABS	ABS		ABS			2700	79
80	1359GL	2635	2639	2697	2910	2930	NI	2960	NI	NI	3106	ABS	ABS	ABS	ABS	ABS	3110	3145	4272	80
81	1650GL	2908	NI	2963	3160?	3175	NI	3205	NI	NI	3359	ABS	ABS	ABS	ABS	ABS	3369	3380	4159	81
82	1444	ABS	2847	2908	3124	3148	NI	3175	3183	3266	3315	ABS	ABS	ABS	ABS	ABS	3329	3355	4599	82
83	1260	2433	2438	2490	2690	2710	NI	2745	2755	2847	2905	ABS	ABS	ABS	ABS	ABS	2915	2933	4158	83
84	1440GL	ABS	2445	2513	2687	2701	NI	2748	2780	2854	2907	ABS	ABS	ABS	ABS	ABS	2930	2946	4142	84
85	1452GL	ABS	2700	2790	2945	2955	NI	3000	NI	NI	3150	ABS	ABS	ABS	3170	ABS	3173	NI	3504	85
86	1521GL	ABS	2835	2960	3120?	3125	NI	3163	3178	NI	3312	ABS	ABS	ABS	3320	ABS	3322	3342	4150	86
87	1400	2654	2662	2752	2893	2908	NI	2958	2990	3058	3112	ABS	ABS	ABS	ABS	ABS	3152	3161	4398	87
88		2796	2802	2889	3020	3033	NI	3086	NI	NI	NI	ABS	ABS	ABS	ABS	ABS	NI	NI	4516	88
89		2817	2825	2911	3034	3047	NI	3100	3122	3196	3248	ABS	ABS	ABS	ABS	ABS	3295	3305	4493	89
90	1535	2758	2766	2852	2970	2984	NI	3038	3058	3126	3183	ABS	ABS	ABS	ABS	ABS	3228	3237	4440	90
91	1590	2817	2827	2913	3030	3044	NI	3097	3118	3186	3243	ABS	ABS	ABS	ABS	ABS	3287	3298	4522	91
92		2770	2784	2868	2980	2994	NI	3047	3067	3141	3193	ABS	ABS	ABS	ABS	ABS	3238	3253	4477	92
93	1590GL	2828	2850	2940	3032	3055	NI	3105	3133			ABS	ABS	ABS		ABS			4525	93
94	1745GL	2896	2915	2999	3100	3112	NI	3172	3199	3272	3334	ABS	ABS	ABS	ABS	ABS	3378	3397	4551	94

Chemung County

1	1637GL	1990	2045	NI	NI	NI	NI	3065	NI	NI	ABS	ABS	ABS	ABS	3107	NR			3163	1
2	1501GL	1963	2025	NI	NI	NI	NI	3050	NI	NI	ABS	ABS	ABS	ABS	3096	NR			3119	2
3	1663GL	2125	2202	NI	NI	NI	NI	3358	NI	NI	ABS	ABS	NI	ABS	3431	NR			3447	3
4	1526GL	2034	2102	2203	2520	3060	3145	3200	3213	3236	ABS	ABS	3243	ABS	3250	3300	NR		3311	4
5	1115GL	1552	1626	1730	2051	2635	2725	2792	2810	2850	ABS	ABS	2855	ABS	2870	2923	NR		2964	5
6	1083GL	NI	1612	NI	NI	NI	NI	2812	NI	NI	ABS	ABS	ABS	ABS	2838	2895	NR		3085	6
7	1080GL	1760	1828	1923	2250	2765	2840	2895	2907	2932	ABS	ABS	2938	ABS	2947	NR			2986	7
8	1227GL	1995	2055	2155	NI	NI	3028	3068	NI	NI	ABS	ABS	ABS	ABS	3114	NR			3134	8
9	1284GL	1834	1903	NI	NI	NI	NI	3160	NI	NI	ABS	ABS	ABS	ABS	3222	3298	NR		3362	9
10	1077GL	1620	1710	1837	2140	2770	2917	2975	2991	NI	ABS	ABS	NI	ABS	3031	3087	3325	3445	11145	10
11	1490GL	2036	2123	2245	2541	3134	3276	3336	3350	3367	ABS	ABS	ABS	ABS	3370	3393?	NR		3401	11
12	1441GL	2017	2105	2225	2520	3110	3245	3309	3322	3338	ABS	ABS	ABS	ABS	3342	3390?	NR		3410	12
13	1275GL	1938	2034	NI	NI	NI	NI	3398	NI	NI	ABS	ABS	NI	ABS	3440	3515	NR		3555	13
14	1680GL	2380	2465	2590	2900	3480	3612	3670	3682	3698	ABS	ABS	ABS	ABS	3703	3750	NR		3824	14
15	1325GL	1910	1990	2110	2393	3057	3130	3207	NI	NI	ABS	AS	3229	ABS	3241	3309	NR		3311	15
16	1418GL	2133	2221	2337	2639	3225	3341	3405	3417	3440	ABS	ABS	3444	ABS	3453	3504	3747	3841	5508	16
17	1670GL	2402	2495	2600	2930	NI	NI	3702	NI	NI	ABS	ABS	3727	ABS	3739	3789	NR		3822	17
18	1605GL	2500	2571	NI	NI	NI	NI	3735	NI	NI	ABS	ABS	ABS	ABS	3775	3845	NR		4015	18
19	849GL	1455	1525	1640	1985	NI	NI	2635	NI	NI	ABS	ABS	ABS	ABS	2678	NR			2697	19
20	911GL	2150	NS	NI	NI	NI	3210	3275	NI	NI	ABS	ABS	ABS	ABS	3312	NR			3433	20
21	1570GL	3125	3195	3300	3635	4110	4185	4250	4263	4282	ABS	ABS	ABS	ABS	4287	NR			4318	21
22	1240GL	2964	3038	3133	3474	3908	3975	4037	4050	4073	ABS	ABS	ABS	ABS	4079	4110	NR		4136	22
23	1290GL	2695	2813	2932	3230	3862	4004	4073	4089	4120	ABS	ABS	4127	ABS	4153	4197	NR		4256	23

Chenango County

1	1373DF			10	350	858	1426	1460	1480	1528	ABS	ABS	ABS	ABS	ABS	1550	1740	1815	5701	1
2	1717GL	1455	1534	1825	2155	2698	3285	3323	3336	3382	ABS	ABS	3392	3415	3434	3452	3688	3767	5520	2
3	1694GL	1782	1869	2150	2483	3063	3611	3643	3665	3705	ABS	ABS	3717	3746	3776	3831	4126	4208	6427	3
4	1400GL	1725	1805	2075	2397	2949	3563	3617	3653	3687	ABS	ABS	3698	3727	3750	3793	4135	4236	6611	4
5	1541GL	1570	1620	1895	2210	2750?	3360	3430	NI	NI	ABS	ABS	3485	ABS	ABS	3520	3775	3880	6292	5
6	1365GL	1294	1360	NS	NS	NS	2970	3034	3050	NI	ABS	ABS	3083	ABS	3131	3155	NR		3254	6
7	1397KB	1370?	1420?	1635	1935	2395	2950	3008	NI	NI	ABS	ABS	NR						3049	7

Well	Elev	TlGb	Mos	Lud	Skan	Mar	ChVl	Sen	MoNd	Edg	BoBl	Scho	CaCn	Esop	OrGl	Held	Rond	Sal	TD	Well
Cortland County																				
1	1509GL	647	667	NI	1175?	NI	NI	1933	NI	NI	ABS	ABS	1982	ABS	ABS	1990	2105	2193	2214	1
2	1481GL	795	819	NI	1260?	NI	2048	2085	2100	NI	ABS	ABS	2138	ABS	2150	2151	NR		2237	2
3	1569GL	1105	1160	1374	1656	2195	2605	2646	2667	2696	ABS	ABS	2713	ABS	2732	2752	2947	3016	8272	3
4	1200GL	1030	1083	1245	1490	NI	2437	2490	NI	NI	ABS	ABS	2539	ABS	ABS	2556	2713	NR	2785	4
5	1168GL	1030	1080	1250	1475	NI	2435	2479	NI	NI	ABS	ABS	2540	ABS	ABS	2570	2710	2800	3400	5
6	1177DF	1036	1073	1260	1500	NI	NI	2502	NI	NI	ABS	ABS	2557	ABS	ABS	2577	2725	2805	3218	6
7	1170GL					NI	2434	2480	2500	2532	ABS	ABS	2539	ABS	ABS	2552	2714	NI	3222	7
Delaware County																				
1	1485GL	1065	1265	1665	2035	2625	3260	3317	3341	3387	ABS	ABS	3410	3445	3494	3515	3774	3838	7952	1
2	1990GL	1680	1890	2275	2650	3240	3884	3946	3973	4020	ABS	ABS	4037	4076	4132	4153	4415	4490	5325	2
3	1457GL	1355	1555	1950	2330	2920	3575	3638	NI	NI	ABS	ABS	3735	3770	3824	3846	NS	4195	4570	3
4	1658GL	1830	2025	2410	2800	3376	3973	4037	4067	4118	ABS	ABS	4135	4166	4218	4246	4514	4590	7958	4
5	1740GL	2747	3010	3400	3740	4393	5055	5145	5170	5234	ABS	5250	5295	5349	5433	5495	5778	5853	10992	5
6	1997GL	1750	2083	2503	2880	3512	4250	4360	4374	4427	ABS	4448	4493	4547	4623	4657	4920	4980	6742	6
7	1834GL	1780	2113	2530	2910	3560	4305	4425	4438	4497	ABS	4519	4565	4625	4705	4750	5010	5070	9075	7
8	1685GL	2425	2703	3110	3453	4151	4905	5050	5074	5137	ABS	5155	5221	5293	5370	5420	5696	5764	6681	8
Erie County																				
1	583GL				10	103	NI	150	166	267	318	ABS	ABS	ABS	ABS	ABS	ABS	328	4313	1
2	684				232	290	NI	337	352	450	ABS	ABS	ABS	ABS	ABS	ABS	ABS	516	1490	2
3	679GL	ABS	150	NI	295	345	NI	390	NI	NI	ABS	ABS	ABS	ABS	ABS	ABS	ABS	568	1423	3
4	683GL					NS	NI	400	NI	NI	ABS	ABS	ABS	ABS	ABS	ABS	ABS	580	1484	4
5	600GL	ABS	160	NI	360	400	NI	441	NI	NI	ABS	ABS	ABS	ABS	ABS	ABS	ABS	630	1474	5
6	585GL	ABS	165	178	372	405	NI	460	475	578	640	ABS	ABS	ABS	ABS	ABS	ABS	650	1545	6
7	702GL	ABS	264	291	478	535	NI	575	593	688	ABS	ABS	ABS	ABS	ABS	ABS	ABS	750	1648	7
8	756GL	ABS	500	NI	NI	765	NI	815	NI	NI	ABS	ABS	ABS	ABS	ABS	ABS	ABS	980	1894	8
9	740GL	ABS	521	548	746	780	NI	828	842	942	ABS	ABS	ABS	ABS	ABS	ABS	ABS	1008	1929	9
10	740GL	ABS	650	675	820	863	NI	910	NI	NI	ABS	ABS	ABS	ABS	ABS	ABS	ABS	1110	2020	10
11	600GL	ABS	371	384	579	610	NI	658	675	784	849	ABS	ABS	ABS	ABS	ABS	ABS	856	1769	11
12	862GL					145	NI	187	198	295	339	ABS	ABS	ABS	ABS	ABS	ABS	340	1344	12
13	886			44	188	326	NI	357	375	465	ABS	ABS	ABS	ABS	ABS	ABS	ABS	510	1471	13
14	840KB	ABS	72	131	286	402	NI	436	454	552	ABS	ABS	ABS	ABS	ABS	ABS	ABS	593	1513	14
15	1010	ABS	211	285	430	565	NI	597	613	710	ABS	ABS	ABS	ABS	ABS	ABS	ABS	751	1754	15
16		ABS	312	355	520	620	NI	653	671	770	ABS	ABS	ABS	ABS	ABS	ABS	ABS	810	1761	16
17	1060					820	NI	849	866	964	ABS	ABS	ABS	ABS	ABS	ABS	ABS	NR	1945	17
18	875GL						NI	710?	730	NI	856	ABS	ABS	ABS	ABS	ABS	ABS	858	1810	18
19	1070KB	ABS	505	580	742	880	NI	910	930	1024	ABS	ABS	ABS	ABS	ABS	ABS	ABS	1075	2165	19
20	1370GL	ABS	967	1015	1193	1275	NI	1305	1323	1427	ABS	ABS	ABS	ABS	ABS	ABS	ABS	1475	2437	20
21	1453DF	ABS	1088	1130	1320	1392	NI	1424	1441	1545	ABS	ABS	ABS	ABS	ABS	ABS	ABS	1593	2525	21
22	1180GL	ABS	965	998	1195	1250	NI	1283	1302	1410	ABS	ABS	ABS	ABS	ABS	ABS	ABS	1457	2404	22
23	1266						NI	1456	1474	1577	ABS	ABS	ABS	ABS	ABS	ABS	ABS	1628	2543	23
24	1560GL	ABS	1458	1503	1705	1767	NI	1810	NI	NI	ABS	ABS	ABS	ABS	ABS	ABS	ABS	1974	2908	24
25	1260GL	ABS	1185	1210	1412	1464	NI	1498	1514	1617	ABS	ABS	ABS	ABS	ABS	ABS	ABS	1669	2634	25
26	1407GL	ABS	924	996	1160	1295	NI	1322	1341	1439	ABS	ABS	ABS	ABS	ABS	ABS	ABS	1485	5541	26
27	1055GL					1036	NI	1060	1080	1190	ABS	ABS	ABS	ABS	ABS	ABS	ABS	1229	2188	27
28	1570GL	ABS	1345	NI	NI	NI	NI	1700?	NI	NI	ABS	ABS	ABS	ABS	ABS	ABS	ABS	1850?	2850	28
29	1067GL	ABS	954	1002	1210	1282	NI	1309	1331	1443	ABS	ABS	ABS	ABS	ABS	ABS	ABS	1492	2594	29
30	1196GL	ABS	1118	1163	1375	1445	NI	1473	1492	1606	ABS	ABS	ABS	ABS	ABS	ABS	ABS	1652	2706	30
31	1636GL	ABS	1607	1645	1861	1923	NI	1950	1967	2068	ABS	ABS	ABS	ABS	ABS	ABS	ABS	2115	3100	31
32	1400GL	ABS	1475	1503	1730	1787	NI	1813	1833	1933	ABS	ABS	ABS	ABS	ABS	ABS	ABS	1985	2984	32
33	1670GL	ABS	1765	1805	1996	2112	NI	2135	2157	2255	ABS	ABS	ABS	ABS	ABS	ABS	ABS	2302	6265	33
34	1600GL	ABS	1710	1751	1940	2060	NI	2082	2102	2200	ABS	ABS	ABS	ABS	ABS	ABS	ABS	2245	3286	34
Genesee County																				
1	860GL						NI	100?	NI	NI	ABS	ABS	ABS	ABS	ABS	ABS	ABS	257	1196	1
2	900GL						NI	290	NI	NI	ABS	ABS	ABS	ABS	ABS	ABS	ABS	450	1473	2
3	1068GL	ABS	52	125	262	414	NI	444	459	555	ABS	ABS	ABS	ABS	ABS	ABS	ABS	592	1569	3
4	930GL				5	163	190	196	211	293	ABS	ABS	ABS	ABS	ABS	ABS	ABS	341	1293	4
5	925KB				35	190	NI	230	NI	NI	ABS	ABS	ABS	ABS	ABS	ABS	ABS	360	1409	5
6	1030GL			30	160	356	380	385	404	479	ABS	ABS	ABS	ABS	ABS	ABS	ABS	533	4340	6
7	940GL				NI	NI	NI	138	NI	NI	290	ABS	ABS	ABS	ABS	ABS	ABS	300	1408	7

Well	Elev	TlGb	Mos	Lud	Skan	Mar	ChVl	Sen	MoNd	Edg	BoBl	Scho	CaCn	Esop	OrGl	Held	Rond	Sal	TD	Well
8	934DF				NI	NI	NI	120	NI	NI	ABS	ABS	ABS	ABS	ABS	ABS	ABS	267	1403	8
9	928GL					NI	NI	247	NI	NI	386	ABS	ABS	ABS	ABS	ABS	ABS	396	1576	9

Greene County

Well	Elev	TlGb	Mos	Lud	Skan	Mar	ChVl	Sen	MoNd	Edg	BoBl	Scho	CaCn	Esop	OrGl	Held	Rond	Sal	TD	Well
1	1928GL	625	930	1358	1730	2400	3515	3718	3723	3810	ABS	3846	3872	3950	4065	4080	4335	4380	7185	1
2	160GL							420	NI	NI	ABS	595	660	NR					705	2

Herkimer County

Well	Elev	TlGb	Mos	Lud	Skan	Mar	ChVl	Sen	MoNd	Edg	BoBl	Scho	CaCn	Esop	OrGl	Held	Rond	Sal	TD	Well
1	1515GL									53	ABS	ABS	73	ABS	ABS	87	263	307	3581	1

Livingston County

Well	Elev	TlGb	Mos	Lud	Skan	Mar	ChVl	Sen	MoNd	Edg	BoBl	Scho	CaCn	Esop	OrGl	Held	Rond	Sal	TD	Well
1	786GL					103	NI	125	150	213	ABS	ABS	ABS	ABS	ABS	ABS	ABS	268	1470	1
2	893GL			15	135	365	385	392	410	475	ABS	ABS	ABS	ABS	ABS	ABS	ABS	530	1720	2
3	881GL			45	175	405	430	435	450	517	ABS	ABS	ABS	ABS	ABS	ABS	ABS	570	5090	3
4	986DF	ABS	170	295	420	NI	NI	695	NI	NI	ABS	ABS	ABS	ABS	ABS	ABS	ABS	838	5648	4
5	1106					773	799	804	822	889	ABS	ABS	ABS	ABS	ABS	ABS	ABS	947	2180	5
6	980GL	ABS	220	345	472	707	730	737	755	825	ABS	ABS	ABS	ABS	ABS	ABS	ABS	880	2077	6
7	1092GL	ABS	305	428	550	783	NI	807	830	900	ABS	ABS	ABS	ABS	ABS	ABS	ABS	957	2137	7
8	1017GL	ABS	245	368	490	725	NI	750	773	840	ABS	ABS	ABS	ABS	ABS	ABS	ABS	897	2081	8
9	730			0	141	377	400	403	NI	NI	ABS	ABS	ABS	ABS	ABS	ABS	ABS	545	1110	9
10	737			44	177	415	NI	450	NI	NI	ABS	ABS	ABS	ABS	ABS	ABS	ABS	594	1150	10
11	951GL	ABS	230	355	480	715	740	747	765	830	ABS	ABS	ABS	ABS	ABS	ABS	ABS	885	2107	11
12	979KB	ABS	280	403	526	760	786	791	810	875	ABS	ABS	ABS	ABS	ABS	ABS	ABS	931	2096	12
13	685GL					NS	NI	32?	NI	115	ABS	ABS	ABS	ABS	ABS	ABS	160	180	1415	13
14	563						15	25	40	100	ABS	ABS	ABS	ABS	ABS	ABS	ABS	150	1400	14
15	563GL					223	250	260	275	334	ABS	ABS	ABS	ABS	ABS	ABS	ABS	385	1678	15
16	579DF			NS	225	NI	NI	505	528	NI	ABS	ABS	ABS	ABS	ABS	ABS	ABS	628	1945	16
17	1096GL	ABS	685	803	924	1162	NI	1190	1212	1272	ABS	ABS	ABS	ABS	ABS	ABS	ABS	1328	2600	17
18	909DF						NI	451	468	530	ABS	ABS	ABS	ABS	ABS	ABS	ABS	580	1901	18
19	1082GL	ABS	365	498	628	886	915	930	946	1015	ABS	ABS	ABS	ABS	ABS	ABS	1065	1090	2517	19
20	1146GL	ABS	529	661	785	1042	1070	1083	1100	1162	ABS	ABS	ABS	ABS	1210	ABS	1212	1220	2625	20
21	594GL			266	390	640	668	683	691	750	ABS	ABS	ABS	ABS	ABS	ABS	ABS	794	2292	21
22	873GL	ABS	515	658	783	1030	1059	1068	1088	1147	ABS	ABS	ABS	ABS	ABS	ABS	1198	1205	6387	22
23	935GL				NI	NI	NI	468	NI	NI	ABS	ABS	ABS	ABS	593	ABS	597	603	1936	23
24	1064GL	ABS	170	312	445	705	725	747	760	830	ABS	ABS	ABS	ABS	ABS	ABS	880	888	2326	24
25	1082	ABS	280	431	570	825	852	865	888	954	ABS	ABS	ABS	ABS	ABS	ABS	1003	1024	1432	25
26	1370GL					1306	1330	1346	1364	1425	ABS	ABS	ABS	ABS	ABS	ABS	1461	1482	3021	26
27	1610GL						1794	1811	1824	1873	ABS	ABS	ABS	ABS	ABS	ABS	1903	1920	3411	27
28	1630GL	ABS	1305	1460	1580	1845	1880	1895	1908	1945	ABS	ABS	ABS	ABS	1980	ABS	1985	2005	3567	28
29	599GL	ABS	372	523	642	900	932	942	961	1010	ABS	ABS	ABS	ABS	1048	ABS	1057	1112	6378	29
30	976GL	ABS	943	1083	1200	1445	1481	1490	1510	1555	ABS	ABS	ABS	ABS	ABS	ABS	ABS	1595	3150	30

Madison County

Well	Elev	TlGb	Mos	Lud	Skan	Mar	ChVl	Sen	MoNd	Edg	BoBl	Scho	CaCn	Esop	OrGl	Held	Rond	Sal	TD	Well
1	1358GL					195	505	526	543	575	ABS	ABS	ABS	ABS	ABS	595	747	812	4927	1
2	1565GL					465	912	940	960	992	ABS	ABS	ABS	ABS	1006	1013	1177	1240	2982	2
3	1494GL					465	929	961	989	1025	ABS	ABS	ABS	ABS	1040	1060	1237	1305	2820	3
4	1640GL			O?	260	NS	NS	1238	NI	NI	ABS	ABS	ABS	ABS	1300	1305	NS	NS	3265	4
5	1635DF			0?	250	770	1215	1243	1264	1295	ABS	ABS	ABS	ABS	1305	1318	1485	1550	3215	5
6	1694			70	353	880	1310	1344	1350	1391	ABS	ABS	ABS	ABS	1399	1402	1578	1648	2789	6
7	1544GL			0	267	795	1233	1265	1285	1313	ABS	ABS	ABS	ABS	1323	1333	1503	1570	5703	7
8	1526GL			100	376	896	1325	1357	1373	1407	ABS	ABS	ABS	ABS	1415	1425	1593	1657	3463	8
9	1416GL					95	540	565	580	622	ABS	ABS	ABS	ABS	ABS	654	835	897	2106	9
10	1605GL					160	694	718	735	786	ABS	ABS	ABS	ABS	ABS	810	993	1060	2214	10
11	1703GL					305	815	845	859	910	ABS	ABS	ABS	ABS	935	947	1130	1195	2352	11
12	1273						460	490	504	550	ABS	ABS	ABS	ABS	ABS	575	755	817	1693	12
13	1526GL					455	883	917	935	969	ABS	ABS	ABS	ABS	987	1000	1170	1237	2539	13
14	1506GL				10	547	1062	1093	1099	1140	ABS	ABS	ABS	ABS	1156	1164	1342	1409	4872	14

Oneida County

Well	Elev	TlGb	Mos	Lud	Skan	Mar	ChVl	Sen	MoNd	Edg	BoBl	Scho	CaCn	Esop	OrGl	Held	Rond	Sal	TD	Well
1	1273GL						355	385	396	447	ABS	ABS	ABS	ABS	ABS	475	653	715	1896	1
2	1319GL					0	460	486	495	545	ABS	ABS	ABS	ABS	575	584	760	823	4366	2

Well	Elev	TlGb	Mos	Lud	Skan	Mar	ChVl	Sen	MoNd	Edg	BoBl	Scho	CaCn	Esop	OrGl	Held	Rond	Sal	TD	Well

Onondaga County

Well	Elev	TlGb	Mos	Lud	Skan	Mar	ChVl	Sen	MoNd	Edg	BoBl	Scho	CaCn	Esop	OrGl	Held	Rond	Sal	TD	Well
1	997GL					347	474	488	512	553	ABS	ABS	ABS	ABS	ABS	561	630	688	4181	1
2	1130GL					435	590	603	626	669	ABS	ABS	ABS	ABS	ABS	680	753	808	2607	2
3	1070GL				0	436	605	617	642	676	ABS	ABS	ABS	ABS	ABS	685	754	806	2763	3
4	955GL					139	324	338	356	400	ABS	ABS	ABS	ABS	ABS	411	495	550	2322	4
5	1370GL				277	728	941	958	979	1010	ABS	ABS	ABS	ABS	ABS	1019	1099	1162	3049	5
6						391	634	654	692	705	ABS	ABS	ABS	ABS	714?	724	850	912	1447	6
7	1827GL	552	580	772	1026	1533	1765	1785	1805	1827	ABS	ABS	ABS	ABS	1835	1843	1956	2017	4171	7

Ontario County

Well	Elev	TlGb	Mos	Lud	Skan	Mar	ChVl	Sen	MoNd	Edg	BoBl	Scho	CaCn	Esop	OrGl	Held	Rond	Sal	TD	Well
1	714GL				195	435	NI	493	503	580	ABS	ABS	ABS	ABS	620	ABS	625	667	2056	1
2	919KB				210	488	512	525	536	606	ABS	ABS	ABS	ABS	ABS	ABS	648?	659?	2122	2
3	725GL							98	NI	NI	ABS	ABS	ABS	ABS	ABS	ABS	219	233	1880	3
4	878GL					140	NI	197	NI	NI	ABS	ABS	ABS	ABS	ABS	320	383	480	1868	4
5	1158DF							705	718	790	ABS	ABS	ABS	ABS	828	ABS	830	855?	2215	5
6	996			112	250	530	547	563	578	647	ABS	ABS	ABS	ABS	ABS	ABS	ABS	690	2094	6
7	785DF					250	NI	303	NI	385	ABS	ABS	ABS	ABS	ABS	ABS	NI	423?	2055	7
8	1050GL	ABS	104	267	414	706	726	743	760	823	ABS	ABS	ABS	ABS	ABS	ABS	ABS	860	2668	8
9	1480DF	ABS	685	840	983	1273	1297	1321	1343	1406	ABS	ABS	ABS	ABS	1450	ABS	1470	1495	2985	9
10	1529GL	ABS	765	922	1065	1355	1375	1394	1417	1475	ABS	ABS	ABS	ABS	1512	ABS	1526	1556	3069	10
11	1330GL					1191	1213	1234	1254	1316	ABS	ABS	ABS	ABS	1348	ABS	1362	1400	2909	11
12	1135DF	ABS	395	547	685	975	995	1019	1041	1101	ABS	ABS	ABS	ABS	1139	ABS	1154	1180	2745	12
13	1008DF					860	880	903	923	985	ABS	ABS	ABS	ABS	1020	ABS	1030	1066	2580	13
14	1101GL					902	925	945	962	1026	ABS	ABS	ABS	ABS	ABS	ABS	1060	1100	2590	14
15	875GL							704	722	785	ABS	ABS	ABS	ABS	ABS	ABS	821	857	2352	15
16	793GL					605	625	650	666	729	ABS	ABS	ABS	ABS	ABS	ABS	764	800	2308	16
17	936DF					800	826	850	869	930	ABS	ABS	ABS	ABS	965	ABS	975	1014	2506	17
18	1202GL	ABS	515	670	810	1097	1120	1140	1161	1220	ABS	ABS	ABS	ABS	1258	ABS			1288	18
19	1042KB	ABS	345	505	640	925	940	966	NI	NI	ABS	ABS	ABS	ABS	1088	ABS	1115	NR	1116	19
20	1735GL	ABS	1330?	NI	1640	NI	NI	1971	NI	NI	ABS	ABS	ABS	ABS	ABS	ABS	2065	2100	3850	20
21	904GL			40	190	445	NI	520	NI	NI	ABS	ABS	ABS	ABS	ABS	ABS	ABS	627	2380	21
22	795GL				20	285	300	320	333	400	ABS	ABS	ABS	ABS	ABS	ABS	ABS	437	2121	22
23	1080GL	137	144	335	495	790	813	827	851	904	ABS	ABS	ABS	ABS	ABS	ABS	ABS	932	6000	23
24	768GL				142	448	477	494	517	571	ABS	ABS	ABS	ABS	ABS	ABS	ABS	598	2435	24
25	858DF				NS	NS	NI	607	NI	NI	ABS	ABS	ABS	ABS	ABS	ABS	ABS	722	2506	25

Otsego County

Well	Elev	TlGb	Mos	Lud	Skan	Mar	ChVl	Sen	MoNd	Edg	BoBl	Scho	CaCn	Esop	OrGl	Held	Rond	Sal	TD	Well
1	1627GL				233	763	1370	1404	1423	1476	ABS	ABS	1496	1527	ABS	1553	1737	1795	2723	1
2	1530GL	83	234	600	915	1435	2035	2061	2092	2139	ABS	ABS	2160	2190	ABS	2210	2395	2460	3689	2
3	1408GL	NS	NS	770?	1100?	1640?	NI	2282	NI	NI	ABS	ABS	2375	2397	ABS	2429	NR		2450	3
4	1156GL	130	285	660	978	1525	2132	2165	2198	2237	ABS	ABS	2260	2300	ABS	2320	2545	2613	3978	4
5	1144GL	1383	1542	1893	2245	2868	3410	3472	3500	3545	ABS	ABS	3559	3590	3620	3656	3922	4010	5835	5
6	1379GL			330	687	1218	1860	1900	1918	1972	ABS	ABS	2002	2034	ABS	2055	2280	2338	5839	6
7	1979GL			0	385	945	1670	1724	1729	1804	ABS	ABS	1830	1860	ABS	1897	2130	2175	5521	7
8	1851GL		200	620	1027	1575	2233	2281	2295	2352	ABS	ABS	2385	2425	2446	2465	2685	2740	3526	8
9	1608GL		60	475	885	1440	2095	2150	2162	2217	ABS	ABS	2250	2285	2313	2325	2550	2608	3406	9
10	1252GL			255	653	1215	1883	1935	1947	2005	ABS	ABS	2039	2074	2105	2108	2350	2402	5118	10

Schoharie County

Well	Elev	TlGb	Mos	Lud	Skan	Mar	ChVl	Sen	MoNd	Edg	BoBl	Scho	CaCn	Esop	OrGl	Held	Rond	Sal	TD	Well
1	2095GL	625	960	1382	1820	2440	3210	3293	3300	3360	ABS	3387	3409	3459	3507	3523	3787	3845	4282	1

Schuyler County

Well	Elev	TlGb	Mos	Lud	Skan	Mar	ChVl	Sen	MoNd	Edg	BoBl	Scho	CaCn	Esop	OrGl	Held	Rond	Sal	TD	Well
1	1511	1200	1224	1455	NI	2047	2092	2117	2141		ABS	ABS	ABS	ABS					2205	1
2	1262	858	881	1093	1286	1649	1700	1725	1749	1787	ABS	ABS	ABS	ABS	1794	1809?			4459	2
3	1228	845	865	1092	1265	1629	1666	1692	1716	1750	ABS	ABS	ABS	ABS	1757	NR			1766	3
4	605GL	413	442	590	869	1336	1405	1445	1470	1504	ABS	ABS	ABS	ABS	1510	1517?	1684	1736	2697	4
5	1387GL	1242	1278	1412	1680	2146	2212	2258	2282	2319	ABS	ABS	ABS	ABS	2326				2430	5
6	1679GL	1660	1678	NI	2075?	NI	NI	2825	NI	NI	ABS	ABS	ABS	ABS	2860	2875	NR		2913	6
7	1476KB	1615	1660	NI	2100	NI	NI	2635	NI	NI	ABS	ABS	ABS	ABS	2680	2690	2830	NI	2890	7
8	1926DF	2342	2390	NI	NI	3305?	NI	3400	NI	NI	ABS	ABS	ABS	ABS	3436	3457	NR		3470	8
9	1870GL	1852	1882	1995	NI	NI	NI	2942	NI	NI	ABS	ABS	ABS	ABS	3020	3035	NR		3037	9
10	1574DF	1543	1573	1685	1995	NI	2640	2682	NI	NI	ABS	ABS	ABS	ABS	2730	2750	NR		2773	10

Well	Elev	TlGb	Mos	Lud	Skan	Mar	ChVl	Sen	MoNd	Edg	BoBl	Scho	CaCn	Esop	OrGl	Held	Rond	Sal	TD	Well
Seneca County																				
1	500GL					62	95	112	135	194	ABS	ABS	ABS	ABS	ABS	ABS	ABS	210	2073	1
2	535GL					132	170	186	210	265	ABS	ABS	ABS	ABS	ABS	ABS	ABS	280	2099	2
3	542DF					35	77	94	116	165	ABS	ABS	ABS	ABS	ABS	ABS	182	190	5537	3
4	469GL							14	36	85	ABS	ABS	ABS	ABS	ABS	ABS	100	115	2286	4
5	552					72	115	130	154	204	ABS	ABS	ABS	ABS	ABS	ABS	220	230	2068	5
6	522GL														50	58	NI	160	1993	6
7	660GL				15	330	365	382	405	455	ABS	ABS	ABS	ABS	ABS	ABS	470	482	2261	7
Steuben County																				
1	2071GL	2713	2741	2913	3018	NI	3317	3333	NI	NI	ABS	ABS	ABS	ABS	ABS	ABS	3378		3550	1
2	2040	2712	2743	2890	3012	3250?	ABS	3274	reef		ABS	ABS	ABS	ABS	3390	ABS	3400		3448	2
3	1765	2435	2466	2614	2733	2995	3029	3045	3065	3102	ABS	ABS	ABS	ABS	ABS	ABS	3111	3161	3184	3
4	1790GL	2738	2777	2919	3045	3306	3332	3348	3361	3403	ABS	ABS	ABS	ABS	3410	3421	3440		3465	4
5	1455GL	1811	1843	2040	2196	2496	2530	2550	2570	2605	ABS	ABS	ABS	ABS	2610	2615?			2663	5
6	1487GL	2030	2063	2245	2335	2673	2703	2723	2743	2757	ABS	ABS	ABS	ABS	2765	NI	2798	2845	4849	6
7	1703GL	2666	2703	2883	NI	NI	3330	3050	NI	NI	ABS	ABS	ABS	ABS	3391	3395	3416		3438	7
8	1865GL	2964	2998	3168	3294	NI	3592	3615	NI	3663	ABS	ABS	ABS	ABS	3670	3690			3692	8
9	1802GL	2897	2933	3110	3245	NI	3530	3552	NI	NI	ABS	ABS	ABS	ABS	3597	NR			3608	9
10	1376GL	2626	2668	2842	2983	NI	3255	3272	NI	NI	ABS	ABS	ABS	ABS	3314	NR			3330	10
11	1984GL	3312	3357	3525	3645	3920	NI	3950	NI	NI	ABS	ABS	ABS	ABS	3985	NR			3993	11
12	1964GL	3320	3373	NI	NI	NI	NI	3970	NI	NS	ABS	ABS	ABS	ABS	4020	NR			4030	12
13	2109KB	3480	3518	NI	NI	4140	4155	4189?	NI	NI	ABS	ABS	ABS	ABS	4211	NR			4223	13
14	1634KB	3050	3095	NI	3415	3700	3720	3735	NI	NI	ABS	ABS	ABS	ABS	3770	3796			3838	14
15	1620GL	3020	3056	3247	3375	3662	3680	3705	3715	3733	ABS	ABS	ABS	ABS	3740	3752?			3788	15
16	1553	2959	3003	3175	3296	3575	3590	3611	3621	3649	ABS	ABS	ABS	ABS	3654	3661			3670	16
17	1865GL	3470	3535?	3725	NI	NI	NI	4138	NI	NI	ABS	ABS	ABS	ABS	4169	NR			4171	17
18	1621GL	3330	3380	NI	NI	NI	NI	3985	NI	NI	ABS	ABS	ABS	ABS	4018	NR			4021	18
19	2344GL	4072	4127	NI	NI	NI	4697	4712	NI	NI	ABS	ABS	ABS	ABS	4756	NR			4777	19
20	2039KB	3808	3850	4025	4153	4421	4450	4480	NI	NI	ABS	ABS	ABS	ABS	4520	4555	4575		4603	20
21	2092	3866	3910	4087	4214	4481	4500	4525	4533	4548	ABS	ABS	ABS	ABS	4560	4570			4614	21
22	1891	3680	3700	NI	NI	NI	NI	4310	NI	NI	ABS	ABS	ABS	ABS	4334	4365			4396	22
23	1858GL	2060	2092	2298	2457	2770	2800	2825	2850	2887	ABS	ABS	ABS	ABS	2900	2907	2975	3025	9808	23
24	1357GL	1341	1365	1588	1745	2070	2103	2126	2150	2183	ABS	ABS	ABS	ABS	2198	2210	NR		2293	24
25	1417GL	1419	1440	1660	1820	2150	2184	2207	2233	2258	ABS	ABS	ABS	ABS	2269	2279	NR		2299	25
26	1174GL	1280	1306	1530	1680	2015	2049	2070	2093	2115	ABS	ABS	ABS	ABS	2123	2133	NR		2230	26
27	738GL	1131	1167	NI	NI	NI	NI	1917	1937	NI	ABS	ABS	ABS	ABS	1962	1976	2040	NR	2059	27
28	1167	802	823	1048	1228	1590	1630	1660	1678	1717	ABS	ABS	ABS	ABS	1722	NR			1737	28
29	1538	2293	2326	2527	2678	3021	3051	3087	3095	3110	ABS	ABS	3115	ABS	3125	3138	NR		3183	29
30	1727KB	2905	2945	NI	NI	NI	NI	3690	NI	NI	ABS	ABS	ABS	ABS	3710	3725	3783	3855	5880	30
31	1630GL	2956	2985	3174	3318	ABS	ABS	3547	reef		ABS	ABS	ABS	ABS	3749	NR			3795	31
32	1780	3062	3090	3277	3410	ABS	ABS	3640	reef		ABS	ABS	ABS	ABS	3845	3875			3885	32
33	1918GL	3420	3465	NI	NI	NI	NI	4010	reef		ABS	ABS	ABS	ABS	NR				4150	33
34	2125GL	3594	3645	NI	NI	NI	4280	4300	NI	NI	ABS	ABS	ABS	ABS	4353	NR			4360	34
35	2122	3575	3609	3793	3939	ABS	ABS	4157	reef		ABS	ABS	ABS	ABS	4335	4373			4435	35
36	2120GL	3725	3765	3950	4086	4368	4381	4412	4417	NI	ABS	ABS	ABS	ABS	4436	NI			4513	36
37	1400GL	3060	3100	NI	NI	NI	NI	3747	NI	NI	ABS	ABS	ABS	ABS	3770	3831			3840	37
38	1355	3146	3180	NI	NI	3780	NI	3812	NI	NI	ABS	ABS	ABS	ABS	3842	NR			3847	38
39	2168DF	3964	4003	4180	NI	NI	NI	4639	NI	NI	ABS	ABS	ABS	ABS	4673	4684			4701	39
40	1667GL	3480	3525	3695	3835	4093	4112	4138	4145	NI	ABS	ABS	ABS	ABS	4166	NR			4176	40
41	1804GL	3576	3620	3845	3987	4264	4290	4315	4328	NI	ABS	ABS	ABS	ABS	4355	NR			4375	41
42	2079GL	3880	3934	NS	NS	NS	NS	4515	NI	NI	ABS	ABS	ABS	ABS	4550	NR			4558	42
43	2069DF	3855	3901	NI	NI	NI	NI	4495	NI	NI	ABS	ABS	ABS	ABS	4535	NR			4541	43
44	2345GL	4148	4197	4365	4503	4756	4776	4800	4811	NI	ABS	ABS	ABS	ABS	4832				4839	44
45	2266	4073	4119	4287	4424	4680	4697	4725	4733	NI	ABS	ABS	ABS	ABS	4757				4767	45
46	2224	4020	4069	4238	4375	4630	4650	4675	4685	NI	ABS	ABS	ABS	ABS	4705	NR			4705	46
47	2190GL	4124	4170	4330	NI	NI	NI	4790	NI	NI	ABS	ABS	ABS	ABS	4833	NR			4840	47
48	2154GL	3990	4020	NI	NI	NI	NI	4690	NI	NI	ABS	ABS	ABS	ABS	4702	4735	NR		4740	48
49	1948GL	3720	3770	NI	NI	NI	NI	4412	NI	NI	ABS	ABS	ABS	ABS	4437	4471	NR		4499	49
50	1761GL	3178	3214	3395	3559	3860	3892	3925	3931	NI	ABS	ABS	ABS	ABS	3947	4000?			3990	50
51	1073GL	1815	1855	NI	NI	NI	NI	2615	NI	NI	ABS	ABS	ABS	ABS	2636	2653	2725	2830	3342	51
52	1636DF	2298	2336	NI	NI	NI	NI	3105	NI	NI	ABS	ABS	3125	ABS	3140	3148	NR		3161	52
53	2190GL	4124	4170	4330	NI	NI	NI	4790	NI	NI	ABS	ABS	ABS	ABS	4833	NR			4840	53
54	1643GL	2115	2155	NI	NI	NI	NI	3041	NI	NI	ABS	ABS	3060	ABS	3078	3101	NR		3109	54
55	1750GL	2304	2345	2560	2730	3120	3162	3206	3216	3225	ABS	ABS	3230	ABS	3236				3254	55
56	1689GL	2307	2350	NI	NI	NI	3207	3247	NI	NI	ABS	ABS	3275	ABS	3281	3308	NR		3323	56

Well	Elev	TlGb	Mos	Lud	Skan	Mar	ChVl	Sen	MoNd	Edg	BoBl	Scho	CaCn	Esop	OrGl	Held	Rond	Sal	TD	Well
57	1060GL	2070	2100	NI	NI	NI	NI	2889	NI	NI	ABS	ABS	ABS	ABS	2897	NR			2918	57
58	1438GL	2460	2503	2697	2877	3240	3285	3323	3330	NI	ABS	ABS	ABS	ABS	3352				3354	58
59	1046GL	2018	2055	NI	NI	NI	2858	2885	NI	NS	ABS	ABS	2910	ABS	2937	NR			2947	59
60	1516GL	2706	2750	2933	3121	3460	3498	3535	3540	NR									3560	60
61	1198GL	2784	2827	3000	3187	3483	3515	3547	3557	3565	ABS	ABS	ABS	ABS	3572	NR			3598	61
62	1807GL	3429	3479	NI	NI	NI	NI	4166	NI	NI	ABS	ABS	ABS	ABS	4193	NR			4202	62
63	1642GL	3322	3373	NI	NI	NI	NI	4108	NI	NI	ABS	ABS	ABS	ABS	4119F?				4180	63
64	1783GL	3408	3463	3625	3797	4088	NI	4132	NI	NI	ABS	ABS	ABS	ABS	4153	NR			4165	64
65	1722GL	3740	3788	NI	NI	NI	NI	4534	NI	NI	ABS	ABS	ABS	ABS	4552	NR			4568	65
66	1770DF	3510	3565	NI	NI	NI	NI	4219	NI	NI	ABS	ABS	ABS	ABS	4235	4280			4282	66
67	2034GL	3907	3949	4115	4278	4537	4556	ABS	4584	NI	ABS	ABS	ABS	ABS	4596	4622			4645	67
68	2254GL	4095	4135	4305	4465	ABS	ABS	4661	reef		ABS	ABS	ABS	ABS	4814				4814	68
69	2218GL	4120	4164	4333	4515	4919	4937	4942	reef		ABS	ABS	ABS	ABS	4994	5030			5055	69
70	2255GL	4140	4185	NI	NI	NI	NI	4820	NI										4852	70
71	2102GL	4005	4048	4217	4378	4730	4748	4770	4775	NI	ABS	ABS	ABS	ABS	4795				4815	71
72	2197GL	4054	4098	4263	4429	4677	4693	4702	reef		ABS	ABS	ABS	ABS	4750	4780			4814	72
73	2184GL	3998	4034	4213	4362	ABS	ABS	4542	reef		ABS	ABS	ABS	ABS	4737	4762			4799	73
74	2269GL	4202	4241	4410	NI	NI	NI	4902	NI	NI	ABS	ABS	ABS	ABS	4921	NR			4931	74
75	2316	4254	4296	4463	4624	4868	4887	4913	4924	NI	ABS	ABS	ABS	ABS	4941	4965			5015	75
76	2260GL	4250	4303	NI	NI	NI	NI	4946	NI	NI	ABS	ABS	ABS	ABS	4970	NR			4980	76
77	2083GL	4132	4185	4335	NI	NI	NI	4795	NI	NI	ABS	ABS	ABS	ABS	4833	NR			4838	77
78	2234GL	NS	NS	NS	NS	NS	NS	4913	NI	NI	ABS	ABS	ABS	ABS	4941	NR			4946	78
79	1245	2915	2965	3145	3355	3650	3678	3713	3725	NI	ABS	ABS	ABS	ABS	3743				3772	79
80	1404GL	3100	3155	3340	NI	NI	NI	3900	NI	NI	ABS	ABS	ABS	ABS	3930	3953	NR		3960	80
81	886	3152	3194	NI	NI	NI	NI	3914	NI	NI	ABS	ABS	3938	ABS	3955	3978			8625	81
82	1645DF	3270	3325	NI	NI	NI	4010	4045	4055	4070	ABS	ABS	ABS	ABS	4075	4105	4210	4252	13500	82
83	1718GL	3186	3235	3410	3615	3870	3910	3942	3951	NI	ABS	ABS	3967	ABS	3974	3999	4100	4165	9790	83
84	1932GL	4140	4185	4360	4595	NI	NI	4945	NI	NI	ABS	ABS	ABS	ABS	4972	4997	NR		5080	84
85	1815GL	4135	4180	4337	4533	F	NI	NR											5130	85
86	2265GL	4557	4622	4780	NI	NI	NI	5247	NI	NI	ABS	ABS	5270	ABS	5277	NR			5318	86
87	2163KB	4398	4441	4590	NI	NI	NI	5080	NI	NI	ABS	ABS	ABS	ABS	5106	NR			5123	87
88	2071DF	4295	4340	4503	4695	4938	4957	4987	5000	5011	ABS	ABS	5013	ABS	5024				5026	88
89	2263GL	4578	4622	NI	NI	NI	NI	5230	NI	NI	ABS	ABS	5255	ABS	5270	NR			5289	89
90	1687GL	2295	2335	NI	NI	3160	3247	3276	NI	NI	ABS	ABS	3300	ABS	3308	NR			3343	90
91	1801DF	2326	2375	NI	NI	NI	3470	3518	NI	NI	ABS	ABS	3545	ABS	3553	3589			3599	91
92	1741GL	2646	2685	NI	NI	NI	3560	3601	NI	NI	ABS	ABS	3630	ABS	3653	NR			3696	92
93	1271GL	2145	2200	NI	NI	3205	3250	3304	NI	NI	ABS	ABS	ABS	ABS	3351	3401			3414	93
94	1794GL	2666	2720	NI	NI	NI	NI	3684	NI	NI	ABS	ABS	ABS	ABS	3746	3791			3794	94
95	1027GL	1995	2048	NI	NI	NI	NI	2977	NI	NI	ABS	ABS	ABS	ABS	3030	NR			3052	95
96	1577GL	2620	2700	NI	NI	NI	NI	3590	NI	NI	ABS	ABS	3607	ABS	3625	NR			3655	96
97	1552KB	2630	2700	NI	NI	NI	3528	3580	NI	NI	ABS	ABS	ABS	ABS	3611	NR			3630	97
98	1402	2510	2563	2693	2955	3365	3424	3468	3480	3493	ABS	ABS	3504	ABS	3520				3582	98
99	1635GL	2925	2975	3165	3355	3725	3765	3805	3816	NI	ABS	ABS	3833	ABS	NI				3864	99
100	1624GL	2867	2905	NI	NI	NI	NI	3789	NI	NI	ABS	ABS	3810	ABS	3816	NR			3838	100
101	1115GL	2437	2486	NI	NI	NI	NI	3356	NI	NI	ABS	ABS	3380	ABS	3395	NR			3436	101
102	1718GL	2912	2968	NI	NI	NI	NI	3780	NI	NI	ABS	ABS	3803	ABS	3817	3880	3960	4070	6825	102
103	1674GL	3016	3063	3177	3500	NI	NI	3990	NI	NI	ABS	ABS	4013	ABS	4028	4055			4063	103
104	1553GL	2854	2910	NS	NS	NS	NI	4007	NI	NI	ABS	ABS	4035	ABS	4070	4106			4214	104

Sullivan County

1	1803GL	4633	4900	5300	5655	6315	7020	7222	7252	7306	ABS	7326	7400	7498	7690	7797	8125	8193	9993	1
2	1560GL	4630	4900	5300	5625	6275	7015	7590	NI	NI	ABS	7745	7880	7990	8240	NR			8275	2

Tioga County

1	965DF	1542	1615	1750	NI	NI	NI	3020	NI	NI	ABS	ABS	ABS	ABS	3044	3106			3150	1
2	1485GL	2720	2792	2985	3270	3915	4260	4337	4355	4376	ABS	ABS	4381	ABS	4397	4455	4705	4804	5154	2
3	1292GL	2724	2834	2990	3253	3863	4110	4185	4200	4227	ABS	ABS	4233	ABS	4268	4405	4779	4913	5228	3
4	980GL	2230	2332	NI	NI	NI	NI	3892	NI	NI	ABS	ABS	3931	ABS	3977	4035	NR		4130	4

Tompkins County

1	590GL	94	120	254	536	1062	1195	1228	1247	1277	ABS	ABS	1283	ABS	ABS	1287	1396	1463	2126	1
2	832	250	285	425	700	1240	1397	1432	1435	1492	ABS	ABS	1500	ABS	1508	1521	1666	1734	2784	2
3	875GL	357	388	515	796	1327	1442	1475	1496	1525	ABS	ABS	1531	ABS	1535	1540	NR		1598	3
4	1455GL	1305	1355	1480	1764	2334	2459	2500	2525	2555	ABS	ABS	2560	ABS	2571	2575	2753	2826	8903	4

Well	Elev	TlGb	Mos	Lud	Skan	Mar	ChVl	Sen	MoNd	Edg	BoBl	Scho	CaCn	Esop	OrGl	Held	Rond	Sal	TD	Well
5	1841GL	1890	1932	2049	2340	2896	2990	3034	3061	3075	ABS	ABS	ABS	ABS	3080	3109	3284	3364	3674	5
6	1041GL	980	1038	1170	1466	2070	2213	2260	2290	2322	ABS	ABS	2327	ABS	2350	2355	2530	2612	9382	6
7	1295DF	1054	1120	1250	1540	2145	2308	2355	2382	2412	ABS	ABS	2420	ABS	2440	2445	2625	2700	10438	7
8	1313GL	1031	1095	1228	1515	2090	2280	2329	2357	2386	ABS	ABS	2391	ABS	2415	2420	2604	2677	8553	8
9	1451DF	1540	1605	NI	NI	NI	NI	2880	NI	NI	ABS	ABS	2930	ABS	2955	2963	NR		3050	9

Ulster County

1	1270GL	2120	2395	2790	3125	3820	4975	5192	5222	5309	ABS	5345	5391	5471	5585	5633	5937	5996	6400	1
2	424GL					503	1902	2405	2418	2525F	ABS	2647	2695	2777	NR				2812	2

Wyoming County

1	1287GL	ABS	427	513	656	822	NI	853	869	NI	ABS	ABS	ABS	ABS	ABS	ABS	ABS	1014	2128	1
2	1442GL	ABS	650	755	885	NI	NI	1055	NI	NI	ABS	ABS	ABS	ABS	ABS	ABS	ABS	1230	2274	2
3	1267GL	ABS	640?	NI	870	1023	NI	1040	NI	NI	ABS	ABS	ABS	ABS	ABS	ABS	ABS	1240	2210	3
4	1502GL	ABS	906	987	1135	1292	NI	1320	1337	1435	ABS	ABS	ABS	ABS	ABS	ABS	ABS	1480	5421	4
5	1163	ABS	285	403	524	739	763	767	785	863	ABS	ABS	ABS	ABS	ABS	ABS	ABS	903	5630	5
6	1501GL	ABS	650	763	890	1100	NI	1120	1145	1228	ABS	ABS	ABS	ABS	ABS	ABS	ABS	1270	6237	6
7	1578GL	ABS	780	893	1020	1230	NI	1252	1274	1356	ABS	ABS	ABS	ABS	ABS	ABS	ABS	1403	5597	7
8	1618	ABS	915	1025	1149	1345	NI	1370	1393	NI	ABS	ABS	ABS	ABS	ABS	ABS	ABS	1523	2680	8
9	1556GL	ABS	779	893	1017	1227	1250	1255	1273	1358	ABS	ABS	ABS	ABS	ABS	ABS	ABS	1402	5644	9
10	1049GL	ABS	306	430	550	746	NI	788	809	876	ABS	ABS	ABS	ABS	ABS	ABS	ABS	930	4975	10
11	1513KB	ABS	700?	NI	NI	NI	NI	1250	NI	NI	ABS	ABS	ABS	ABS	ABS	ABS	ABS	1370	5320	11
12	967GL	ABS	300	422	520	725	NI	750	NI	NI	ABS	ABS	ABS	ABS	ABS	ABS	ABS	890	4944	12
13	1650GL	ABS	1061	1160	1290	1471	NI	1497	1520	1610	ABS	ABS	ABS	ABS	ABS	ABS	ABS	1658	2780	13
14	1609DF	ABS	1074	1165	1300	NI	NI	1539	NI	NI	ABS	ABS	ABS	ABS	ABS	ABS	ABS	1686	5722	14
15	1622GL	ABS	1025	1133	1255	1456	1483	1488	1504	1570	ABS	ABS	ABS	ABS	ABS	ABS	ABS	1627	5503	15
16	1757	ABS	1190	1280	1414	1588	NI	1612	1631	1726	ABS	ABS	ABS	ABS	ABS	ABS	ABS	1777	2861	16
17	1620KB	ABS	1300?	1430	NI	NI	NI	1700	NI	NI	ABS	ABS	ABS	ABS	ABS	ABS	ABS	1910?	2900	17
18	1375KB	ABS	1200?	1280	NI	NI	NI	1630	NI	NI	ABS	ABS	ABS	ABS	ABS	ABS	ABS	1810	2850	18
19	1425	ABS	1288	1341	1492	1653	NI	1678	1699	1786	ABS	ABS	ABS	ABS	ABS	ABS	ABS	1837	2935	19
20	1525KB	ABS	1400	NI	NI	NI	NI	1850	NI	NI	ABS	ABS	ABS	ABS	ABS	ABS	ABS	1986	3125	20
21	1790	ABS	1691	1746	1885	2061	NI	2087	2108	2195	ABS	ABS	ABS	ABS	ABS	ABS	ABS	2246	3412	21
22	1800GL	ABS	1813	1867	2015	2180	NI	2206	2227	2313	ABS	ABS	ABS	ABS	ABS	ABS	ABS	2366	3498	22
23	1760KB	ABS	1800?	1880	NI	NI	NI	2200	NI	NI	ABS	ABS	ABS	ABS	ABS	ABS	ABS	2330	3485	23
24	1830GL	ABS	1858	1911	2052	2232	2257	2259	2281	2355	ABS	ABS	ABS	ABS	ABS	ABS	ABS	2415	3538	24
25	1681GL	ABS	1712	1764	1908	2082	2108	2116	2138	2208	ABS	ABS	ABS	ABS	ABS	ABS	ABS	2259	3427	25
26	1702KB	ABS	1990	2060	NI	NI	NI	2390	NI	NI	ABS	ABS	ABS	ABS	ABS	ABS	ABS	2560	6700	26
27	1051	ABS	345	468	590	820	NI	850	871	940	ABS	ABS	ABS	ABS	ABS	ABS	ABS	995	2192	27
18	1081DF	ABS	394	515	637	872	NI	896	920	985	ABS	ABS	ABS	ABS	ABS	ABS	ABS	1040	2205	28
29	1155GL				NS	915	NI	938	963	1035	ABS	ABS	ABS	ABS	ABS	ABS	ABS	1093	2272	29
30	1260GL					NS	NI	NS	1077	1152	ABS	ABS	ABS	ABS	ABS	ABS	ABS	1206	2379	30
31						992	NI	1017	1037	1106	ABS	ABS	ABS	ABS	ABS	ABS	ABS	1160	2369	31
32	1222GL			NS	NS	1000	1027	1031	1047	1116	ABS	ABS	ABS	ABS	ABS	ABS	ABS	1169	2333	32
33						1080	1107	1111	1126	1196	ABS	ABS	ABS	ABS	ABS	ABS	ABS	1250	2458	33
34	1501GI	ABS	897	1021	1140	1360	NI	1386	1408	1476	ABS	ABS	ABS	ABS	ABS	ABS	ABS	1533	5718	34
35	1708GL	ABS	1586	1682	1805	2008	2032	2037	2052	2120	2177	ABS	ABS	ABS	ABS	ABS	ABS	2190	3442	35
36	1635GL	ABS	1350	1450	1575	1780	1806	1810	1826	1895	ABS	ABS	ABS	ABS	ABS	ABS	ABS	1955	3228	36
37	1768GL	ABS	1580	1680	1797	NI	NI	2055	2070	NI	ABS	ABS	ABS	ABS	ABS	ABS	ABS	2184	3410	37
38	1472GL	ABS	1245	1348	1470	NI	NI	1723	NI	NI	ABS	ABS	ABS	ABS	ABS	ABS	ABS	1855	2431	38
39	1400GL	ABS	1080	1175	NS	NS	NI	1551	NI	NI	ABS	ABS	ABS	ABS	ABS	ABS	ABS	1685	2901	39
40	1345GL	ABS	1170	1300	1433	NI	NI	1683	1700	NI	ABS	ABS	ABS	ABS	ABS	ABS	ABS	1814	2361	40
41	1329	ABS	1157	1275	1395	1619	1647	1655	1668	1725	ABS	ABS	ABS	ABS	ABS	ABS	ABS	1784	2368	41
42	1560GL	ABS	1528	1638	1761	1983	NI	2007	2036	2094	ABS	ABS	ABS	ABS	ABS	ABS	ABS	2150	7182	42
43	1617GL	ABS	1451	1557	1684	1903	1930	1935	1950	2012	ABS	ABS	ABS	ABS	ABS	ABS	ABS	2070	3334	43
44	1592GL	1685	1715	1812	1933	2137	NI	2172	NI	NI	ABS	ABS	ABS	ABS	ABS	ABS	ABS	2307	3598	44
45	1610GL	1653	1660	1740	1869	2075	1998	2103	2122	2178	ABS	ABS	ABS	ABS	ABS	ABS	ABS	2249	3511	45
46	1582	1780	1816	1886	2024	2229	2255	2262	2281	2326	ABS	ABS	ABS	ABS	ABS	ABS	ABS	2398	3718	46
47		ABS	1038	1164	1284	1513	1537	1547	1568	1622	ABS	ABS	ABS	ABS	ABS	ABS	ABS	1676	3020	47

Yates County

1	828GL	95	107	300	505	815	843	866	884	927	ABS	ABS	ABS	ABS	ABS	ABS	ABS	939	2757	1
2	884	262	275	475	675	985	1012	1032	1051	1093	ABS	ABS	ABS	ABS	ABS	ABS	ABS	1105	2896	2
3	916GL	344	350	550	740	NI	NI	1117	NI	NI	ABS	ABS	ABS	ABS	ABS	ABS	ABS	1190	2971	3
4	960GL	325	337	543	746	1059	1085	1105	1123	1161	ABS	ABS	ABS	ABS	ABS	ABS	ABS	1166	3061	4

Well	Elev	TlGb	Mos	Lud	Skan	Mar	ChVl	Sen	MoNd	Edg	BoBl	Scho	CaCn	Esop	OrGl	Held	Rond	Sal	TD	Well
5	724GL	250	270	NI	NI	NI	NI	1118	NI	NI	ABS	ABS	ABS	ABS	1188	NS			1348	5
6	792GL	335	353	532	755	1135	1175	1200	1220	1258	ABS	ABS	ABS	ABS	1264				1434	6
7	733	260	275	445	675	1065	1107	1130	1150	1187	ABS	ABS	ABS	ABS	1193	1215	1270	1322	1377	7
8	1081GL	496	502	NI	NI	NI	1305	1330	NI	NI	ABS	ABS	ABS	ABS	1383	1403			1414	8
9	1476GL	1187	1215	NI	NI	NI	NI	2044	NI	NI	ABS	ABS	ABS	ABS	2093	NR			2106	9

Commonwealth of Pennsylvania

Allegheny County

Well	Elev	TlGb	Mos	Lud	Skan	Mar	ChVl	Sen	MoNd	Edg	BoBl	Scho	CaCn	Esop	OrGl	Held	Rond	Sal	TD	Well
1		6954	7002	7073	7244	7327	7380	ABS	7412	7430	7455	ABS	ABS	ABS	7610				7904	1

Armstrong County

Well	Elev	TlGb	Mos	Lud	Skan	Mar	ChVl	Sen	MoNd	Edg	BoBl	Scho	CaCn	Esop	OrGl	Held	Rond	Sal	TD	Well
1	1462GL	5944	6037	6056	6287	6345	6369	ABS	6390	6410	6419	ABS	ABS	ABS	ABS	6514				1
2	1367GL	6680	6770	6810	7044	7104	7183	ABS	7215	7235	7250	ABS	ABS	ABS	7380				7477	2

Beaver County

Well	Elev	TlGb	Mos	Lud	Skan	Mar	ChVl	Sen	MoNd	Edg	BoBl	Scho	CaCn	Esop	OrGl	Held	Rond	Sal	TD	Well
1	948GL	4113	NI	4125	4170	4180	NI	ABS	4270	4295	4340	ABS	ABS	ABS					6666	1
2	1109GL	4383	4388	4403	4443	4449	NI	ABS	4530	4553	4597	ABS	ABS	ABS					4806	2
3	983GL	4243	4255	4270	NI	NI	NI	ABS	4405	NI	4470	ABS	ABS	ABS	4599	4670	NI	4930	6823	3
4	715GL	4980	4987	5050	5140	5185	NI	ABS	ABS	5222	5259	ABS	ABS	ABS	5385				5443	4

Blair County

Well	Elev	TlGb	Mos	Lud	Skan	Mar	ChVl	Sen	MoNd	Edg	BoBl	Scho	CaCn	Esop	OrGl	Held	Rond	Sal	TD	Well
1	1472GL	4675	ABS	4705	5288	5498	5565	ABS	5675	5687	5695	ABS	ABS	ABS	5730				5823	1

Bradford County

Well	Elev	TlGb	Mos	Lud	Skan	Mar	ChVl	Sen	MoNd	Edg	BoBl	Scho	CaCn	Esop	OrGl	Held	Rond	Sal	TD	Well
1	1599GL	3515	3589	3675	4080	NI	NS	4705?	NS	NS	ABS	ABS	ABS	ABS	4729?				4750	1
2	1520GL	3015	3115	3220	3580	NI	NI	4388	NI	NI	ABS	ABS	4428	ABS	4445	4488			4522	2
3	1350GL						NS	4220	NI	NI	ABS	4262	4277	ABS	4322	NI	4665	4805	7220	3
4	1856GL	4376	4465	NI	4935?	NI	NI	5600	NI	NI	ABS	ABS	5618	ABS	5645	NR			5700	4
5	1445GL	4315?	4405?	NS	NS	NS	NS	5480?	NS	NS	ABS	5510?							5530	5
6	1245GL	4915	5076	NR															5450	6
7	925GL	2582	2680	NI	3265	NI	NI	4382	NI	NI	ABS	NR							4420	7
8	1323GL	3027	3100	NI	NI	NI	NI	4926	NI	NI	ABS	4970	4990?	5060	5140	5230	NR		5290	8
9	1278GL	4693	4800	5110	5403	6083	6420	6570	6581	6610	ABS	6613	6635?	6725	6822	6922	NI	NI	7830	9
10	1372GL	4370	4485	4795	5077	5750	6055	6210	6220	6247	ABS	6253	6290	6385	6466	NR			6610	10
11	705GL	3542	3667	3987	4233	4900	5197	5335	5355	5382	ABS	5387	5410	5505	5588	5738	6054	6194	6899	11
12	761GL	3316	3443	NI	NI	NI	4980	5130	NI	NI	ABS	5188	5210?	NR					5307	12
13	1280GL	3200	3348	NI	3860	NI	4790	4935	NI	NI	ABS	4990	5020	5095	5164	NR			5227	13
14	1120DF	2186	2315	NI	NI	NI	NI	3809	NI	NI	ABS	3860	3885	3960	4033	NR			4104	14
15	1430DF	2846	3035	3230	3553	4135	4250	4374	4390	4414	ABS	4420	4443	4490	4543	4673			4678	15
16	1353GL	4892	5016	5410	5625	NI	NI	6938?	NI	NI	ABS	NI	NI	7270	7504	NR			7507	16
17	1260GL	4225	4343	NI	NI	NI	6100	6262	NI	NR									6290	17
18	1561GL	5853	5975	6373	6604	7271	7525	7680	7700	7745	ABS	7750	7777	7873	8034	8150	8500	8633	12843	18

Butler County

Well	Elev	TlGb	Mos	Lud	Skan	Mar	ChVl	Sen	MoNd	Edg	BoBl	Scho	CaCn	Esop	OrGl	Held	Rond	Sal	TD	Well
1	1306GL	4483	NI	4525	NI	NI	NI	ABS	4692	4730	4770	ABS	ABS	ABS	4828	4841	NI	4983	10096	1
2	1120DF	5090	5120	5170	NI	NI	NI	ABS	5360	NI	5415	ABS	ABS	ABS	5480	5492	NI	5670	7620	2
3	1315GL	5120	5125	NI	NI	NI	NI	ABS	5355	NI	NI	ABS	ABS	ABS	5503	5543	NI	5700	7603	3

Cambria County

Well	Elev	TlGb	Mos	Lud	Skan	Mar	ChVl	Sen	MoNd	Edg	BoBl	Scho	CaCn	Esop	OrGl	Held	Rond	Sal	TD	Well
1	1900GL	7209	ABS	7241	7722	7920	7970	ABS	8045	8057	NR								8063	1
2	1952GL	7260	ABS	7297	7805	NI	8030	ABS	8113	NI	8140	ABS	ABS	ABS	8200	8212			8225	2
3	2056GL	7460	ABS	7485	8003	8170	8233	ABS	8303	8320	8335	ABS	ABS	ABS	8393	NR			8475	3
4	1939GL	7370	ABS	7395	7900	8060	8125	ABS	8202	8217	8230	ABS	ABS	ABS	8293	8315?			8338	4

Well	Elev	TlGb	Mos	Lud	Skan	Mar	ChVl	Sen	MoNd	Edg	BoBl	Scho	CaCn	Esop	OrGl	Held	Rond	Sal	TD	Well
5	2026GL	7480	ABS	7500	8044	8220	8267	ABS	8350	8356	8375	ABS	ABS	ABS	8439	8475?			8514	5
6	2119GL	7885	ABS	7905	NI	NI	NI	ABS	8730	NI	8762	ABS	ABS	ABS	8851	8880			8950	6
7	2439GL	7000	ABS	7020	NI	NI	NI	ABS	7778	NI	7798	ABS	ABS	ABS	7895	7945			8000	7
8	2257GL	6835	ABS	6845	NI	NI	NI	ABS	7570	NI	7580	ABS	ABS	ABS	7672	NR			7675	8

Cameron County

Well	Elev	TlGb	Mos	Lud	Skan	Mar	ChVl	Sen	MoNd	Edg	BoBl	Scho	CaCn	Esop	OrGl	Held	Rond	Sal	TD	Well
1	1968GL	6077	6137	6281	6504	6638	6657	ABS	6702	6707	ABS	ABS	ABS	ABS	ABS	6715			6770	1
2	1187GL	4491	4555	NI	NI	NI	NI	ABS	5091	NI	5100	ABS	ABS	ABS	5120	5121	NR		5181	2
3	1765DF	6120	6170	NI	NI	NI	NI	ABS	6740	NI	ABS	ABS	ABS	ABS	6753	NR			6825	3
4	944GL	5242	5330	5450	5690	5823	5852	ABS	5900	NI	ABS	ABS	ABS	ABS	5920				5982	4
5	1560GL	6132	6217	6335	6553	6670	6694	ABS	6740	6747	6754	ABS	ABS	ABS	6773	NR			6813	5
6	1006GL	4696	4770	NI	NI	NI	NI	ABS	5378	NI	ABS	ABS	ABS	ABS	5388	NR			5406	6
7	970DF	5410	5500	NI	5880	NI	NI	ABS	6105	NI	6120	ABS	ABS	ABS	6129	NR			6151	7
8	1765DF	5792	5920	NI	NI	NI	NI	ABS	6592	NI	6608	ABS	ABS	ABS	6612	6624	NR		6688	8
9	1265GL	5370	5493	NI	NI	NI	NI	ABS	6095	NI	6115	ABS	ABS	ABS	6120	6150	NR		6158	9
10	899GL	5141	5280	NI	NI	NI	NI	ABS	5857	NI	5880	ABS	ABS	ABS	5895	5912	NR		5922	10
11	2175GL	6623	6780	NI	NI	NI	NI	ABS	7392	NI	7415	ABS	ABS	ABS	7432	NR			7438	11
12	2110GL	6612	6700	NI	7100	NI	NI	ABS	7422	NS	ABS	ABS	ABS	ABS	7455	7475	NR		7484	12

Centre County

Well	Elev	TlGb	Mos	Lud	Skan	Mar	ChVl	Sen	MoNd	Edg	BoBl	Scho	CaCn	Esop	OrGl	Held	Rond	Sal	TD	Well
1	2166GL	7255	7475	NI	NI	NI	NI	ABS	8320	NI	ABS	ABS	ABS	ABS	8348	NR			8399	1
2	1669DF	7125	7275	NI	NI	NI	NI	ABS	8180	NI	ABS	ABS	ABS	ABS	8209	NR			8257	2

Clarion County

Well	Elev	TlGb	Mos	Lud	Skan	Mar	ChVl	Sen	MoNd	Edg	BoBl	Scho	CaCn	Esop	OrGl	Held	Rond	Sal	TD	Well
1	1624GL	4877	4910	4967	5137	5143	5180	ABS	5185	5206	5240	ABS	ABS	ABS	ABS	5260	5300	5370	6452	1
2	1481GL	4938	4970	4995	5180	5191	NI	ABS	5230	NI	5270	ABS	ABS	ABS	5308	5325			5409	2
3	1510GL	5130	5165	5212	5373	5377	5412	ABS	5425	5440	5460	ABS	ABS	ABS	5500	5518	5568		7460	3
4	1635GL	4892	4945	4983	NI	NI	NI	ABS	5198	NI	5250?	ABS	ABS	ABS	5262	5275	NI	5375	6782	4
5	1436GL	5210	5250	NI	NI	NI	NI	ABS	5518	NI	5550	ABS	ABS	ABS	5593	5598			6110	5
6	1451KB	5417	5475	5522	5670	5736	5750	ABS	5762	5780	5790	ABS	ABS	ABS	5830	5853	5903	5973	7906	6
7		5118	5155	5187	5343	5397	5419	ABS	5430	5448	5460	ABS	ABS	ABS	ABS	5508	5576	5650	5820	7
8	1363GL	5838	ABS	5910	6130	6200	6224	ABS	6240	6258	6266	ABS	ABS	ABS	ABS	6325			6400	8

Clearfield County

Well	Elev	TlGb	Mos	Lud	Skan	Mar	ChVl	Sen	MoNd	Edg	BoBl	Scho	CaCn	Esop	OrGl	Held	Rond	Sal	TD	Well
1	1599GL	6170	6300?	6323	NI	NI	NI	ABS	NI	NI	NI								7077	1
2	1691GL	6930	7020	7050	NI	NI	NI	ABS	7568	NI	7581	ABS	ABS	ABS	NR				7582	2
3	1459GL	6584	ABS	6686	7005	7094	7117	ABS	7148	7161	7170	ABS	ABS	ABS	7226				7259	3
4	1408DF	6559	ABS	6690	NI	NI	NI	ABS	7170	NI	7186	ABS	ABS	ABS	7255				7270	4
5	2137GL	6297	6437	NI	NI	NI	NI	ABS	7016	NI	7035	ABS	ABS	ABS	7073	7080			7094	5
6	1557DF	6682	ABS	6873	7210	7383	7412	ABS	ABS	7470	7482	ABS	ABS	ABS	7532				7590	6
7	1964GL	6685	ABS	6865	NI	NI	NI	ABS	7428	NI	7440	ABS	ABS	ABS	7500	NI			8049	7
8	1979GL	6605	6730	NI	7130	NI	NI	ABS	7400	NI	NI	ABS	ABS	ABS	7442	7462			7565	8
9	1425GL	6870	7048	7085	7400	7617	7650	ABS	7718	7726	7735	ABS	ABS	ABS	7761	7781			7838	9
10	1668GL	6837	ABS	6948	7392	7617	7658	ABS	7737	7747	7757	ABS	ABS	ABS	7805?				7798	10
11	1681GL	7043	7140	NI	NI	NI	NI	ABS	7990	NI	8010	ABS	ABS	ABS	8032	NI			8085	11
12	1638GL	7123	ABS	7200	7670	7893	7935	ABS	8020	8027	8038	ABS	ABS	ABS	8090				8131	12
13	1567GL	7047	ABS	7107	7565	7818	7860	ABS	7946	7956	7968	ABS	ABS	ABS	8020	8045?			151	13
14	1587DF	7353	ABS	7400	7867	8035	8070	ABS	8133	NI	8150	ABS	ABS	ABS	8235	8300?			8353	14

Clinton County

Well	Elev	TlGb	Mos	Lud	Skan	Mar	ChVl	Sen	MoNd	Edg	BoBl	Scho	CaCn	Esop	OrGl	Held	Rond	Sal	TD	Well
1	1338GL	5670	5755	NI	NI	NI	NI	ABS	6518	NI	ABS	ABS	ABS	ABS	6532				6542	1
2	1757GL	5645	5757	5913	6197	6387	6413	ABS	6500	6507	ABS	ABS	ABS	ABS	6517	6540			6626	2
3	1108GL	5017	5125	5280	5550	NI	NI	ABS	5810	NI	ABS	ABS	ABS	ABS	5819	5824			5836	3
4	1559GL	5494	5623	NI	NS	NS	NS	ABS	6272	NI	ABS	ABS	ABS	ABS	6296	6317?			6317	4
5	1877GL	5499	5580	NI	NI	NI	NI	ABS	6326	NI	ABS	ABS	ABS	ABS	6351				6370	5
6	1723GL	6040	NS	NS	NS	NS	NS	ABS	6856	NI	ABS	ABS	ABS	ABS	6879	6925			6929	6
7	1737GL	6889	7027	NI	NI	NI	NI	ABS	7760	NI	ABS	ABS	ABS	ABS	7785	7807?			7834	7
8	1713GL	6410	6520	NI	NI	NI	NI	ABS	7384	NI	ABS	ABS	ABS	ABS	7410	7432			7630	8
9	2254GL	6906	7015	NI	NI	NI	NI	ABS	7893	NI	ABS	ABS	ABS	ABS	7916				7942	9
10	1681GL	7240	7390	NI	7860?	NI	NI	ABS	8388	NI	ABS	ABS	ABS	ABS	8420	8525			8576	10
11	2111GL	7076	7195	NI	NI	NI	NI	ABS	8094	NI	ABS	ABS	ABS	ABS	8144	8230			8311	11
12	1804DF	7315	7485	NI	7940?	NI	NI	ABS	8373	NI	ABS	ABS	ABS	ABS	8408	8450			8467	12

Well	Elev	TlGb	Mos	Lud	Skan	Mar	ChVl	Sen	MoNd	Edg	BoBl	Scho	CaCn	Esop	OrGl	Held	Rond	Sal	TD	Well
Columbia County																				
1	890GL	3765	3915	4440	4650	5327	5770	6195	6213	6287	ABS	6293	6365	6418	6458				7445	1
Crawford County																				
1	983DF	1952	NI	ABS	NI	1983	NI	2122	NI	NI	2320	ABS	ABS	ABS	2390	ABS	2416	2470	3541	1
2		1880	NI	ABS	NI	1905	NI	2055	NI	NI	2250	ABS	ABS	ABS	2318	ABS	2324	2375	3463	2
3	1022	1980	1990	ABS	NI	2003	2143	2145	2170	2281	2322	ABS	ABS	ABS	ABS	ABS	2396	2442		3
4	1148GL	2165	NI	ABS	NI	NI	NI	2318	NI	NI	2475	ABS	ABS	ABS	2526	ABS	2536	2595	3585	4
5	1345DF	2428	2481	ABS	NI	NI	NI	2620	NI	NI	2775	ABS	ABS	ABS	2846	ABS	2850		3034	5
6	1243DF	2365	2430	ABS	NI	NI	NI	2559	NI	NI	2710	ABS	ABS	ABS	2744	ABS	2751	2780	3959	6
7	1258GL	2468	2495	ABS	NI	2505	2642	2649	2677			ABS	ABS	ABS		ABS			4093	7
8		2092	2112	ABS	NI	2127	NI	2258	NI	NI	2430?	ABS	ABS	ABS	2468	ABS	2476	2510	3600	8
9	1117GL	2255	NI	ABS	NI	2265	2406	2410	2430			ABS	ABS	ABS		ABS			3772	9
10	1120DF	2240	NI	ABS	NS	NS	NI	2384	NI	NI	NI	ABS	ABS	ABS	2592	ABS	2605	2690	3939	10
11	1100GL	2390	NI	ABS	NI	2402	2540	2546	2566			ABS	ABS	ABS		ABS			4077	11
12	1640GL	3043	3113	3153	3179	3186	3270	3277	3298	3378	3426	ABS	ABS	ABS	ABS	ABS	3467	3479	7435	12
13	1220GL	2635	2705	NI	NI	NI	NI	2836	NI	NI	2990	ABS	ABS	ABS	ABS	ABS	3018	3050	4354	13
14	1303KB	2618	2685	ABS	NI	2713	NI	2820	2840	2923	2972	ABS	ABS	ABS	ABS	ABS	3040	3048		14
15	1342GL	3083	NI	3115	3122	3127	NI	3243	NI	NI	NI	ABS	ABS	ABS	3432	ABS	3447	3470	4644	15
16	1744GL	3646	3693	3735	NI	NI	NI	3880?	NI	NI	4032	ABS	ABS	ABS	4043	ABS	4046?	4085	5322	16
17	1650GL	3603	3657	3717	3785	3793	NI	3852	3866	3915	3961	ABS	ABS	ABS	ABS	ABS	3997	4040	5432	17
18	1622GL	3557	3602	3653	3698	3708	NI	3777	3797	3850	3897	ABS	ABS	ABS	ABS	ABS	3935?	3995		18
Elk County																				
1	1836GL	4872	4878					ABS	5294	NI	5330	ABS	ABS	ABS	ABS	5348			7930	1
2	1572GL	5127	5151	NI	NI	NI	NI	ABS	5546	NI	5565	ABS	ABS	ABS	ABS	5594	NI	5720	7546	2
3	2088KB																		4000	3
4	1903GL	5790	5850	NI	NI	NI	NI	ABS	6305	NI	NI	ABS	ABS	ABS	ABS	6350			6462	4
5	2343GL	5815	5860	5895	NI	NI	NI	ABS	6392	NI	6415	ABS	ABS	ABS	6491	6495	NI	6660	6790	5
6	1457GL	6108	6225	NI	NI	NI	NI	ABS	6750	NI	NI	ABS	ABS	ABS	6773				7000	6
7	2055GL	6208	NS	NS	NS	NI	NI	ABS	6909	NI	NI	ABS	ABS	ABS	6945	NR			6975	7
8	1294GL	6315	6420	6490	NI	NI	NI	ABS	6902	NI	6923	ABS	ABS	ABS	6361	NR			7020	8
9	1781GL	6350	6470	6520?	NI	NI	NI	ABS	7022	NI	7033	ABS	ABS	ABS	7067	7100			7110	9
10	1279DF	5814	5930	5980?	6300	NI	NI	ABS	6485	NI	6497	ABS	ABS	ABS	6536	6566			6588	10
Erie County																				
1	537GL	1100	1115	ABS	NI	1130	NI	1272	NI	NI	1480	ABS	ABS	ABS	1562	ABS	1582	1660	5098	1
2	604DF	830	873	897	923	930	1050	1068	1080	1202	1250	ABS	ABS	ABS	ABS	ABS	1305	1340	4741	2
3		1175	1225	1245	1255	1263	1382	1393	1417	1539	1589	ABS	ABS	ABS	ABS	ABS	1660	1693	2534	3
4	859	1462	1524	1545	1550	1556	1695	1705	1730	1850	1902	ABS	ABS	ABS	1980	ABS	1993	2041		4
5	1250	1887	1940	1970	1975	1983	2110	2115	2140	2252	2300	ABS	ABS	ABS	2358	ABS	2372	NR		5
6	995GL	1645	1710	1733	1740	1743	NI	1872	NI	NI	2070	ABS	ABS	ABS	2145	ABS	2180		2192	6
7	842	1500	1533	ABS	NI	1547	NI	1689	1713	1840	1888	ABS	ABS	ABS	ABS	ABS	1968	2017		7
8	726	1368	1382	ABS	NI	1403	1550	1560	1597	1715	1765	ABS	ABS	ABS	ABS	ABS	1855	1918		8
9	753	1457	1463	ABS	NI	1482	1635	1642	1678	1800	1850	ABS	ABS	ABS	ABS	ABS	1923	2007		9
10	820GL	1560	1570	ABS	NI	1582	NI	1725	NI	NI	1910	ABS	ABS	ABS	ABS	ABS	1998	2050	3065	10
11		1647	1680	ABS	NI	1695	NI	1836	1859	1983	2025	ABS	ABS	ABS	2110	ABS	2122	2157	3123	11
12		1637	NI	ABS	NI	NI	NI	1800	NI	NI	2000	ABS	ABS	ABS	2087	ABS	2100	2150	3265	12
13	883DF	1640	1645	ABS	NI	1663	1803	1806	1830	1952	1998	ABS	ABS	ABS	ABS	ABS	2081	2140		13
14	965GL	1860	1865	ABS	1874	1881	2020	2023	2046	2163	2210	ABS	ABS	ABS	2287	ABS	2295	2336	5935	14
15	1360GL	2180	2250	NI	NI	NI	NI	2413	NI	NI	2600	ABS	ABS	ABS	2677	ABS	2696	NI	3768	15
16	1290GL	2168	2220	ABS	2235	2243	NI	2375	2398	NI	NI	ABS	ABS	ABS		ABS			3765	16
17	878DF						NI	2133	NI	NI	2315	ABS	ABS	ABS	2391	ABS	2413		3471	17
18	1473DF	2462	2520	ABS	2535	2541	2665	2670	2695	2803	2852	ABS	ABS	ABS	2927	ABS	2936	2970	4103	18
19	1413DF	2493	2560	NI	NI	NI	NI	2705	NI	NI	2900	ABS	ABS	ABS	2952	ABS	2965		6840	19
20	1555GL	2480	2535	2562	2610	2615	2700	2708	2732	2826	2872	ABS	ABS	ABS	ABS	ABS	2925	2958	4010	20
21	1463GL	2300	2363	2387	2420	2430	2523	2528	2555	2658	2715	ABS	ABS	ABS	ABS	ABS	2780	2815	7465	21
22	1450GL	2353	2421	2446	2475	2485	2576	2585	2610	2710	2759	ABS	ABS	ABS	ABS	ABS	2827	2887	3993	22
23	1528GL	2508	2575	2603	2650	2660	2745	2750	2770	2872	2919	ABS	ABS	ABS	ABS	ABS	2980	3018	4139	23
24	1760DF	2880	2945	2985	3040	3050	NI	3125	3150?	3240	3300	ABS	ABS	ABS	ABS	ABS	3332	3353	4506	24
25	1440	2688	2757	2800	2846	2856	NI	2933	2960	3040	3097	ABS	ABS	ABS	ABS	ABS	3152	3167		25
Forest County																				
1	1830	4786	4804	NI	NI	NI	NI	ABS	5130	NI	5181	ABS	ABS	ABS	5192	5201			5349	1

Well	Elev	TlGb	Mos	Lud	Skan	Mar	ChVl	Sen	MoNd	Edg	BoBl	Scho	CaCn	Esop	OrGl	Held	Rond	Sal	TD	Well
2	1653GL	4740	4770	4795	5020	5030	NI	ABS	5075	NI	5135	ABS	ABS	ABS	ABS	5153			6855	2
3	1605GL	4487	4510	4539	4763	4768	4793	ABS	4800	4821	4849	ABS	ABS	ABS	ABS	4857	4884	4941	6026	3
4	1550GL	4480	4500	4530	NI	NI	NI	ABS	4750	NI	ABS	ABS	ABS	ABS	ABS	4805	NI	4910	6484	4

Indiana County

1	1967GL	6780	ABS	6870	NI	NI	NI	ABS	7448	NI	7457	ABS	ABS	ABS	7520	7538			7555	1
2	1634GL	6800	ABS	NI	NI	NI	NI	ABS	7425	NI	7440	ABS	ABS	ABS	7567	7580			7795	2
3	1235GL	6760	6860	6905	7147	7215	7307	ABS	7340	7358	7368	ABS	ABS	ABS	7460	7480				3
4	1980GL	7290	ABS	7312	7800	7930	7970	ABS	8030	8043		ABS	ABS	ABS					8199	4
5	1825DF	7374	ABS	7437	NI	NI	NI	ABS	8043	NI	8050	ABS	ABS	ABS	8155				8166	5
6	1468GL	7052	ABS	7090	7548	7655	NI	ABS	7810	7825	7832	ABS	ABS	ABS					7975	6
7	2014GL	7375	ABS	7410	NI	NI	NI	ABS	8120	NI	8135	ABS	ABS	ABS	8232	NR			8232	7
8	1620GL	7073	ABS	7090	7590	7695	7735	ABS	7795	7815	7833	ABS	ABS	ABS	7910	NR				8

Jefferson County

1		5162	ABS	5205	5475	5537	5555	ABS	5572	5587	5593	ABS	ABS	ABS	ABS	5637	NI	5704	7711	1
2	1496GL	NS	ABS	NS	NS	NS	NS	ABS	6080	NI	6098	ABS	ABS	ABS	ABS	6155	NI	6280	7156	2
3	1799DF	6550	ABS	6610	NI	NI	NI	ABS	7030	NI	7044	ABS	ABS	ABS	7106	7122			7139	3
4	1539GL	6836	ABS	6963	7270	7348	7360	ABS	7371	7389	7395	ABS	ABS	ABS	7497?				7544	4
5	1390GL	6610	ABS	6740	NI	NI	NI	ABS	7181	NI	7205	ABS	ABS	ABS	7274	NR			7294	5
6	1953GL	6832	ABS	6993	7320	NI	NI	ABS	7516	NI	7528	ABS	ABS	ABS	7618	7628			7633	6

Lackawanna County

1	1031GL	5200	5364	5960	6235	6862	7255	7458	7504	7622	ABS	7680	7837	8000	8191	8297	NR		8623	1

Luzerne County

1	1236GL	4508	4677	NR															4757	1
2	1030GL	2260	2400	NI	NI	NI	4050	4255	NI	NI	ABS	4375	4430	4565	4707	4790			5555	2
3	1070GL	3400	3550	NI	NI	NI	F												7583	3

Lycoming County

1	2073GL	6845	6962	7125	7407	7680	7710	ABS	7810	7820	ABS	ABS	7828	ABS	7876?				7900	1
2	1578DF	5678	5800	6000	6273	NI	NI	ABS	6768	NI	ABS	ABS	6790	ABS	6818	6847?			6870	2
3	1767DF	5453	5595	5785	6021	6395	6430	6542	6545	NI	ABS	6558	6570	ABS	6595				6638	3
4	1547DF	6208	6335	6560	6710	NI	NI	7408	NI	NI	ABS	7425	NI	ABS	7465	7570			7579	4
5	1788GL	6590	6725	6950	7100	7587	7640	7770	7778	7795	ABS	7800	7817	ABS	7841	7930?			7990	5
6	1556GL	5860	5995	6217	6350	6855	6917	7040	7050	NI	ABS	7065		ABS					7220	6
7	1276DF	5095	5222	5507	5670	6309	6393	6536	6541	6560	ABS	6569	6588	ABS	6643	6684	6937	7071	8118	7
8	1254GL	5429	NS	NI	NI	NS	NS	NR											6854	8

McKean County

1	1474GL	ABS	3260	3323	3592	3604	NI	3627	3639	3677	ABS	ABS	ABS	ABS	3712	ABS	3737	3777	5064	1
2	2082	ABS	3975	4029	NI	NI	NI	4338	NI	NI	ABS	ABS	ABS	ABS	4416	ABS	4431	NR	4448	2
3	2090	ABS	3961	4033	NI	NI	NI	4323	NI	NI	ABS	ABS	ABS	ABS	4398	NR			4406	3
4	1358	ABS	3409	3478	NI	NI	NI	3770	NI	NI	ABS	ABS	ABS	ABS	3835	NR			3845	4
5	1603	ABS	3766	3812	NI	NI	NI	4166	NI	NI	ABS	ABS	ABS	ABS	4231	NR			4241	5
6	1730GL	ABS	3900	3967	4261	4266	4286	4295	4300	4330	ABS	ABS	ABS	ABS	4347	NR			4380	6
7	2086GL	ABS	4230	4295	NI	NI	NI	4621	NI	NI	4670	ABS	ABS	ABS	4690	4718			4722	7
8	1531GL	ABS	3700?	3780	NI	NI	NI	4123	NI	NI	ABS	ABS	ABS	ABS	4185	4207	NI	4270	4297	8
9	2137	ABS	4439	4499	NI	NI	NI	4816	NI	NI	ABS	ABS	ABS	ABS	4874	NR			4898	9
10	2040GL	NS	NS	NS	NS	4915	4935	4946	NI	NI	ABS	ABS	ABS	ABS	4993	5020	NI	5086	6940	10
11	1342	NS	NS	NI	NI	NI	NI	4070	NI	NI	ABS	ABS	ABS	ABS	4117	4143	NI	NR	4146	11
12	1550GL	ABS	4240	4315	4600?	NI	NI	ABS	4655	NI	4705	ABS	ABS	ABS	ABS	4720	NI	4830?	8002	12
13	2240GL	ABS	4630	4700	5005	5027	5048	5057	5067	5095	ABS	ABS	ABS	ABS	5112	5125	5137	5185	11891	13
14	2257KB	ABS	4728	4797	5097	5131	5150	5160	5164	5197	ABS	ABS	ABS	ABS	5215	5230	5242	5290	10478	14
15	1555	3993	4040	4158	NI	NI	NI	4526	NI	NI	ABS	ABS	ABS	ABS	ABS	4579	NI		4701	15
16	1488GL	3935	3955	4040	NI	NI	NI	ABS	4445	NI	4473	ABS	ABS	ABS	4498	4502	NI	4565	6615	16
17	1751	4202	4220	NI	NI	NI	NI	ABS	4678	NI	4705	ABS	ABS	ABS	ABS	4726	NI		5135	17
18	1525GL	4203	4218	4310	4592	4641	4667	ABS	4695	4715	ABS	ABS	ABS	ABS	ABS	4728	4747	4795	4896	18
19	1750	4654	4658	NI	NI	NI	NI	ABS	5126	NI	5161	ABS	ABS	ABS	ABS	5182			5269	19
20	2201GL	ABS	5170	5280	NI	NI	NI	ABS	5648	NI	5670	ABS	ABS	ABS	ABS	NR			6787	20

Mercer County

Well	Elev	TlGb	Mos	Lud	Skan	Mar	ChVl	Sen	MoNd	Edg	BoBl	Scho	CaCn	Esop	OrGl	Held	Rond	Sal	TD	Well
1	1160GL	2930	NI	2939	2953	2959	3073	3078	3100	3171	3219	ABS	ABS	ABS	3285	ABS	3295	3370	4791	1
2	1275GL	3045	NI	NI	NI	NI	NI	3186	NI	NI	3330	ABS	ABS	ABS	3418	3430	NI	3505	4961	2
3	950GL	2921	NI	ABS	2929	2935	3053	3060	3080	3144	3185	ABS	ABS	ABS	3250	3270	3300	3370	8210	3
4	905GL	3120	NI	ABS	NI	NI	NI	3247	NI	NI	3360	ABS	ABS	ABS	3435	3465	NI	3580	5127	4
5	1026GL	3269	NI	ABS	3277	3283	NI	3390	3408	3473	3515	ABS	ABS	ABS	3595	3608	3667	3743	5289	5
6	1325GL	3290	3300	3315	3340	3350	NI	3450	NI	NI	3590	ABS	ABS	ABS	3632	3712	NI	3745	9246	6
7	1334GL	3700	3705	3715	3747	3753	NI	3860	3872	3920	3961	ABS	ABS	ABS	4013	4040	4060	4140	9919	7
8	1287GL	3633	3640	3650	3745	3750	NI	3857	NI	NI	3950	ABS	ABS	ABS	3988	4010	NI	4110	5586	8
9	1575GL	4037	4045	4071	4108	4113	NI	ABS	4214	4250	4290	ABS	ABS	ABS	4336	4346	4372	4454	6140	9

Northumberland County

Well	Elev	TlGb	Mos	Lud	Skan	Mar	ChVl	Sen	MoNd	Edg	BoBl	Scho	CaCn	Esop	OrGl	Held	Rond	Sal	TD	Well
1	704GL	673	750	NI	NI	2140	2450	2684	NI	NI	ABS	2760							5804	1
2	780DF					360	675	1060	NI	NI	ABS	1125							1819	2

Pike County

Well	Elev	TlGb	Mos	Lud	Skan	Mar	ChVl	Sen	MoNd	Edg	BoBl	Scho	CaCn	Esop	OrGl	Held	Rond	Sal	TD	Well
1	1043GL	3410	3695	4075	4450	5165	6085	6403	6432	6530	ABS	6550	6646	6896	7310	7420	NR		7563	1
2	1484GL	4945	5190	5650	6000	NI	7605	7865	NI	NI	ABS	8060	8150	8400	8570	8650	9085	9200?	13910	2
3	1575GL	5645	5868	NR							ABS								5990	3

Potter County

Well	Elev	TlGb	Mos	Lud	Skan	Mar	ChVl	Sen	MoNd	Edg	BoBl	Scho	CaCn	Esop	OrGl	Held	Rond	Sal	TD	Well
1	1833GL	4073	4125	4215	4423	NI	NI	4601	4618	NI	ABS	ABS	ABS	ABS	4653				4673	1
2	2127GL	4387	4435	NI	NI	NI	NI	4909	NI	NI	ABS	ABS	ABS	ABS	4960				4980	2
3	2150GL	4210	4270	NI	NI	NI	NI	4802	NI	NI	ABS	ABS	ABS	ABS	4878				4889	3
4	2200GL	4323	4380	4513	4697	4847	NI	4880	NI	NI	ABS	ABS	ABS	ABS	4916				4923	4
5	2155GL	4292	4346	NI	NI	NI	NI	4850	NI	NI	ABS	ABS	ABS	ABS	4880?				4911	5
6	2086GL	4150	4205	4335	NI	NI	NI	4738	NI	NI	ABS	ABS	ABS	ABS	4772				4793	6
7	1923GL	4053	4097	NI	NI	NI	NI	4722	NI	NI	ABS	ABS	ABS	ABS	4744				4755	7
8	2506GL	5064	5112	NI	NI	NI	NI	5765	NI	NI	ABS	ABS	ABS	ABS	5794	5824			5857	8
9	2106GL	4585	4637	4795	5033	5160	5175	ABS	5210	5228	ABS	ABS	ABS	ABS	5233				5269	9
10		4794	4832	NI	NI	NI	NI	ABS	5694	NI	ABS	ABS	5712	ABS	5732	NR			5744	10
11	2088GL	4562	4608	NI	NI	NI	NI	ABS	5192	NI	ABS	ABS	ABS	ABS	5209				5218	11
12	2230GL	4718	4765	NI	NI	NI	NI	ABS	5305	NI	ABS	ABS	ABS	ABS	5329				5353	12
13	2238GL	4905	4972	5095	NI	NI	NI	ABS	5501	NI	ABS	ABS	ABS	ABS	5516				5529	13
14	2180	4567	4615	NI	NI	NI	NI	ABS	5147	NI	ABS	ABS	ABS	ABS	ABS	5174			5561	14
15	2296GL	5440	5500	5636	5855	5944	5965	ABS	6003	6017	ABS	ABS	ABS	ABS	ABS	6030	6055	6103	6155	15
16	1933GL	4438	4510	NI	NI	4948	NI	ABS	4994	NI	5020	ABS	ABS	ABS	5044	5046	NI	5135	8482	16
17	2027GL	4780	4834	5025	NI	NI	NI	ABS	NR			ABS		ABS					6001	17
18	2072GL	4710	4770	4933	NI	NI	NI	ABS	5440	NI	ABS	ABS	ABS	ABS	5468	5469			5609	18
19	2318GL	4383	4436	NI	NI	NI	NI	ABS	5029	NI	ABS	ABS	ABS	ABS	5044				5218	19
20		4800	4853	NI	NI	NI	NI	ABS	5487	NI	ABS	ABS	ABS	ABS	ABS	5520?			5685	20
21	2158GL	5472	5537	NI	NI	NI	NI	ABS	6289	NI	ABS	ABS	NI	ABS	ABS	6342			6454	21
22	1880GL	5273	5333	NI	NI	NI	NI	ABS	6104	NI	ABS	ABS	NI	ABS	6170	6171			6314	22
23	2441GL	6007	6071	6233	6470	6690	6710	ABS	6780	6795	ABS	ABS	6800	ABS	6840	6844	NR		6909	23
24	1941GL	5476	5526	NI	NI	NI	NI	ABS	6295	NI	ABS	ABS	ABS	ABS	6313	6334			6463	24
25	2196GL	6157	6217	6383	6597	6798	6818	ABS	6876	6895	ABS	ABS	6903	ABS	6936	6943	7047	7113	7298	25
26	2306DF	6304	6383	NI	NI	NI	NI	ABS	7000	NI	ABS	ABS	ABS	ABS	7022	7045			7081	26
27	1337GL	5489	5545	5704	5910	6073	6090	ABS	6140	6151	ABS	ABS	ABS	ABS	6162	6172	6197		6211	27
28	2393GL	5860	5925	6070	NI	NI	NI	ABS	6477	NI	ABS	ABS	ABS	ABS	6495	6544			6572	28
29	2110GL	6156	6240	6410	6693	6922	6955	ABS	7038	7054	ABS	ABS	ABS	ABS	7060				7096	29
30	2166GL	5882	5952	NI	NI	6610	NI	ABS	6722	NI	ABS	ABS	ABS	ABS	6746				6774	30
31	1689DF	6020										ABS		ABS	NR				6040	31
32	1908GL	5484	5564	NI	NI	NI	NI	ABS	6204	NI	ABS	ABS	ABS	ABS	6217				6230	32
33	2120	5566	5654	NI	NI	NI	NI	ABS	6283	NI	ABS	ABS	ABS	ABS	6302				6314	33
34	1847GL	5550	5640	5800	6075	6265	6295	ABS	6365	6379	ABS	ABS	ABS	ABS	6385	6420	6569	6644	18834	34

Sullivan County

Well	Elev	TlGb	Mos	Lud	Skan	Mar	ChVl	Sen	MoNd	Edg	BoBl	Scho	CaCn	Esop	OrGl	Held	Rond	Sal	TD	Well
1	1743DF	5230	5340	5740	5950	6600	6713	6875	6888	6918	ABS	6925	6950	7025	7090	NR			7192	1
2	1478DF	6700	6847	7203	7425	8020	8190	8368	8375	8413	ABS	8420	8468	8548	8665	8755	8995	9129	12343	2

Susquehanna County

Well	Elev	TlGb	Mos	Lud	Skan	Mar	ChVl	Sen	MoNd	Edg	BoBl	Scho	CaCn	Esop	OrGl	Held	Rond	Sal	TD	Well
1	1447GL	3265	3380	3660	4030	4680	NI	5313	NI	NI	ABS	5360	5390	5440	5527	NR			5577	1

Well	Elev	TlGb	Mos	Lud	Skan	Mar	ChVl	Sen	MoNd	Edg	BoBl	Scho	CaCn	Esop	OrGl	Held	Rond	Sal	TD	Well
2	1655GL	3475	3590	3880	4241	4905	5405	5535	5545	5575	ABS	5580	5607	5666	5756	NR			5874	2
3	1232GL	4426	4530	4830	5480	6110	NI	6848	NI	NI	ABS	6915	7000	7170	7280	7390?	7875	8000	8534	3

Tioga County

Well	Elev	TlGb	Mos	Lud	Skan	Mar	ChVl	Sen	MoNd	Edg	BoBl	Scho	CaCn	Esop	OrGl	Held	Rond	Sal	TD	Well
1	1853GL	4430	4475	4653	4867	5127	5153	5188	5202	5210	ABS	ABS	5212	ABS	5225	5239	5344	5400	5515	1
2	2145GL	4353	4400	NI	4753	NI	NI	5081	NI	NI	ABS	ABS	5100	ABS	5112				5140	2
3	1474	2909	2991	NI	NI	NI	NI	3913	NI	NI	ABS	ABS	ABS	ABS	3930	3980			7148	3
4	1764GL	3034	3120	NI	3553	NI	NI	ABS	3985	NI	ABS	ABS	ABS	ABS	4005	4047			4086	4
5	1713GL	4200	4260	NI	NI	NI	NI	ABS	5140	NI	ABS	ABS	NI	ABS	5172	5215			5217	5
6	1680GL	3099	3173	NI	NI	NI	NI	ABS	3999	NI	ABS	ABS	4018	ABS	4030				4037	6
7	1917GL	4067	4133	4270	4570	4834	4875	ABS	4929	NI	ABS	ABS	4942	ABS	4950	4973?			5003	7
8	1844DF	3686	3740	3900	NI	NI	NI	ABS	4472	NI	ABS	ABS	ABS	ABS	4475	4501			4505	8
9	1941GL	3821	3872	NI	NI	NS	NS	ABS	4587	NI	ABS	ABS	ABS	ABS	4598				4628	9
10	1660GL	4425	4475	4635	4850	5070	5095	ABS	5146	NI	ABS	ABS	5158	ABS	5166	5175			5195	10
11	2248GL	NR																	3400	11
12	1335GL	4075	4179	NI	NI	NI	NI	ABS	5258	NI	ABS	ABS	ABS	ABS	5282	5346			5684	12
13	1415GL	4544	4631	NI	NI	NI	NI	ABS	5505	NI	ABS	ABS	ABS	ABS	5520	5560			5639	13
14	1780GL	4555	4618	NI	NI	NI	NI	ABS	5515	NI	ABS	ABS	ABS	ABS	5529	5580			5714	14
15	1836GL	4512	4606	NI	5045	NI	NI	ABS	5492	NI	ABS	ABS	ABS	ABS	5516	5580			5663	15
16	1875DF	4990	5080	NI	NI	NI	NI	ABS	5930	NI	ABS	ABS	ABS	ABS	5948	5995			6081	16
17	1569GL	4762	4838	4970	5281	5520	5545	ABS	5628	5642	ABS	ABS	ABS	ABS	5646	5665	5830	5893	9091	17
18	1447GL	4512	4596	4720	5030	NI	NI	ABS	5376	NI	ABS	ABS	ABS	ABS	5393	5418			5424	18
19	1677DF	4565	4640	NI	NI	NI	NI	ABS	5425	NI	ABS	ABS	ABS	ABS	5450	5471			5477	19
20	1327GL	6253	6358	6468	6735	7100	7140	ABS	7245	NI	ABS	ABS	ABS	ABS	7259	7290			10300	20
21	2390GL	6370	6444	NI	NI	NI	NI	ABS	7264	NI	ABS	ABS	7285	ABS	7294				7316	21

Venango County

Well	Elev	TlGb	Mos	Lud	Skan	Mar	ChVl	Sen	MoNd	Edg	BoBl	Scho	CaCn	Esop	OrGl	Held	Rond	Sal	TD	Well
1	1058DF	3560	NI	NI	NI	NI	NI	ABS	3792	NI	3880	ABS	ABS	ABS	3912				3915	1
2	1424GL	4073	4120	4170	4257	4262	NI	ABS	4330	4365	4408	ABS	ABS	ABS	ABS	4450	4470	4540	6085	2
3	1474GL	4295	4307	4357	4429	4433	NI	ABS	4519	4544	4586	ABS	ABS	ABS	4648	4650	4690	4774	6510	3
4	1462GL	4320	4328	4380	4445	4455	NI	ABS	4540			ABS	ABS	ABS					6635	4
5	1420GL	4376	4381	4430	4512	4520	NI	ABS	4599	NI	4665	ABS	ABS	ABS	4730	4732	NI	4870	6690	5

Warren County

Well	Elev	TlGb	Mos	Lud	Skan	Mar	ChVl	Sen	MoNd	Edg	BoBl	Scho	CaCn	Esop	OrGl	Held	Rond	Sal	TD	Well
1	1360DF	ABS	2740?	2850	2990?	2995	NI	3046	NI	NI	3180	ABS	ABS	ABS	3200	ABS	3210		4435	1
2	1525GL	2940	2980	3082	3144	3152	NI	3220	3240	3308	3352	ABS	ABS	ABS	ABS	ABS	3382	3422	4653	2
3	1620GL	ABS	3300	3405	3570	3590	NI	3625	3640	3688	3741	ABS	ABS	ABS	ABS	ABS	3759	3779	8130	3
4	1262DF	3150	3170	3305	NI	NI	NI	3475	NI	NI	3580	ABS	ABS	ABS	3595	ABS	3598		4898	4
5	1849	3471	3530	3626	NI	NI	NI	3752	NI	NI	3883	ABS	ABS	ABS	ABS	ABS	3907	3963	5165	5
6	1745GL	3595	3645	3705	3780?	3790	NI	3872	NI	NI	3990	ABS	ABS	ABS	ABS	ABS	4015	4065	5316	6
7	1649DF	3830	NI	3925	NI	NI	NI	4125	NI	NI	ABS	ABS	ABS	ABS	4203	ABS	4218	4285	5595	7
8	1266	ABS	3086	3148	NI	NI	NI	3444	NI	NI	ABS	ABS	ABS	ABS	3526	ABS	3538	NR	3542	8
9	1689	ABS	3515	3588	3842	3852	NI	3880				ABS	ABS	ABS					4000	9
10	1926	ABS	3760	3830	NI	NI	NI	4104	NI	NI	ABS	ABS	ABS	ABS	4197	4199	4257	4287	5630	10
11	2019GL	ABS	4073	4163	4380	4390	NI	4413				ABS	ABS	ABS						11
12	2055	ABS	4375	4502	NI	NI	NI	4745	NI	NI	ABS	ABS	ABS	ABS	4799	4819	NR		4830	12
13	1994	ABS	4323	4427	NI	NI	NI	4666	NI	NI	ABS	ABS	ABS	ABS	4711	4738	4778	4825	4928	13
14	1944	ABS	4281	4395	NI	NI	NI	4620	NI	NI	ABS	ABS	ABS	ABS	4679	NR			4689	14
15	1945GL	ABS	4330	4440	4638	4643	4666	4671	4677	4704	ABS	ABS	ABS	ABS	4726	4744	4762	4816	5681	15
16	1714GL	4173	4182	4215	4420	4428	NI	ABS	4463	4485	4523	ABS	ABS	ABS	4531	ABS	4545	4613	6038	16
17	1583GL	3985	NI	4036	NI	NI	NI	ABS	4279	NI	ABS	ABS	ABS	ABS	4354	ABS	4366	4443	5850	17
18	1708DF	4153	NI	4210	NI	NI	NI	ABS	4450	NI	ABS	ABS	ABS	ABS	4528	ABS	4545	4620	9410	18
19	1627DF	4075	4105	4145	4330	4340	NI	ABS	4375	4410	4452	ABS	ABS	ABS	4465	ABS	4497	4563	6004	19

Washington County

Well	Elev	TlGb	Mos	Lud	Skan	Mar	ChVl	Sen	MoNd	Edg	BoBl	Scho	CaCn	Esop	OrGl	Held	Rond	Sal	TD	Well
1	1136GL	5722	5733	5790	5937	5963	NI	ABS	6010	6018	6041	ABS	ABS	ABS	6196	6275?			6576	1

Wayne County

Well	Elev	TlGb	Mos	Lud	Skan	Mar	ChVl	Sen	MoNd	Edg	BoBl	Scho	CaCn	Esop	OrGl	Held	Rond	Sal	TD	Well
1	897DF	4830	5055	5450	5830	6580	7100	7480	7527	7587	ABS	7619	7743	7938	8105	8192	8640	8705?	8731	1
2	1489GL	5550?	5700?	NI	NI	NI	7800	7960	NI	NI	ABS	8100	8255	8500	8660	8740?	9200	9300	12250	2

Westmoreland County

Well	Elev	TlGb	Mos	Lud	Skan	Mar	ChVl	Sen	MoNd	Edg	BoBl	Scho	CaCn	Esop	OrGl	Held	Rond	Sal	TD	Well
1		6690	6723	6823	7055	7100	7150	ABS	7190											1
2	1209GL	6619	6669	6757	6940	6995	7052	ABS	7087	7101	7124	ABS	ABS	ABS	7255				7310	2
3	1255GL	6947	ABS	6969	7348	7392	7446	ABS	7493	7512	7529	ABS	ABS	ABS	7663				7740	3

Wyoming County

Well	Elev	TlGb	Mos	Lud	Skan	Mar	ChVl	Sen	MoNd	Edg	BoBl	Scho	CaCn	Esop	OrGl	Held	Rond	Sal	TD	Well
1	960GL	5879	6000	6555	6800	7400?	7800	7946	NI	NI	ABS	8082	8130	8310	8444	NR			8593	1

State of Ohio

Ashtabula County

Well	Elev	TlGb	Mos	Lud	Skan	Mar	ChVl	Sen	MoNd	Edg	BoBl	Scho	CaCn	Esop	OrGl	Held	Rond	Sal	TD	Well
1	785GL	ABS	ABS	ABS	ABS	1538	1654	1670	1692	1804	1850	ABS	ABS	ABS	ABS	ABS	1953	2010	3213	1
2	580GL	ABS	ABS	ABS	ABS	1180	NI	1330	NI	NI	1545	ABS	ABS	ABS	1630	ABS	1660	1715	5320	2
3	636DF	ABS	ABS	ABS	ABS	1260	1377	1390	1419	1529	1579	ABS	ABS	ABS	1675	ABS	1697	1745	2781	3
4		1779	NI	ABS	NI	1785	1917	1925	1944	2060	2108	ABS	ABS	ABS	ABS	ABS	2210	2250	3419	4
5	1088DF						NI	2245	2270	2370	2413	ABS	ABS	ABS	ABS	ABS	2472	2511		5
6	1052DF	2120	NI	ABS	NI	2133	2262	2267	2290	2390	2433	ABS	ABS	ABS	ABS	ABS	2503	2555		6
7	1054KB	ABS	ABS	ABS	ABS	2075	2210	2215	2240			ABS	ABS	ABS					6367	7
8		ABS	ABS	ABS	ABS							ABS	ABS	ABS	ABS	ABS				8
9	1063KB	2312	NI	ABS	NI	2323	2453	2459	2487	2572	2618	ABS	ABS	ABS	ABS	ABS	2680	2743		9
10	1064KB	ABS	ABS	ABS	ABS	2060	NI	2205	2222			ABS	ABS	ABS					3970	10

Columbiana County

Well	Elev	TlGb	Mos	Lud	Skan	Mar	ChVl	Sen	MoNd	Edg	BoBl	Scho	CaCn	Esop	OrGl	Held	Rond	Sal	TD	Well
1	1163KB	3133	NI	ABS	NI	3140	3250	3258	3277	3347	3395	ABS	ABS	ABS	ABS	3565	3689	3782		1
2	1180GL	3703	NI	ABS	NI	3720	NI	3860	3880	3935	3980	ABS	ABS	ABS	4163	4190	4325	4400		2
3	1210GL	3490	NI	3498	3507	3512	NI	3638	3652	3719	3767	ABS	ABS	ABS	3950	3984	4107	4195	5925	3
4		3498	NI	ABS	NI	3504	3632	3635	3654			ABS	ABS	ABS						4
5	1170KB	3529	NI	3534	3547	3555	NI	3689	3705	3764	3810	ABS	ABS	ABS	4015	4053	4180	4260		5
6	1137KB	4023	NI	4031	4058	4063	NI	ABS	4188	4237	4285	ABS	ABS	ABS	4471	4534	4653	4740		6

Coshocton County

Well	Elev	TlGb	Mos	Lud	Skan	Mar	ChVl	Sen	MoNd	Edg	BoBl	Scho	CaCn	Esop	OrGl	Held	Rond	Sal	TD	Well
1								2480	2495	2577?	2610	ABS	ABS	ABS	ABS	ABS	2647	2757		1

Cuyahoga County

Well	Elev	TlGb	Mos	Lud	Skan	Mar	ChVl	Sen	MoNd	Edg	BoBl	Scho	CaCn	Esop	OrGl	Held	Rond	Sal	TD	Well
1							1180	1195	1240	1350	1397	ABS	ABS	ABS	ABS	ABS	1526	1603	1997	1
2	772GL	ABS	ABS	ABS	ABS	1107	1208	1217	1240	1326	1366	ABS	ABS	ABS	ABS	ABS	1460	1535	2723	2
3	1140GL	ABS	ABS	ABS	ABS	1710	1855	1875	1916	2050	2095	ABS	ABS	ABS	ABS	ABS	2215	2320	3645	3
4	1188GL	ABS	ABS	ABS	ABS	1697	1826	1835	1876	1985	2025	ABS	ABS	ABS					3521	4

Lake County

Well	Elev	TlGb	Mos	Lud	Skan	Mar	ChVl	Sen	MoNd	Edg	BoBl	Scho	CaCn	Esop	OrGl	Held	Rond	Sal	TD	Well
1	701	ABS	ABS	ABS	ABS	1110	1277	1308	1350	1488	1535	ABS	ABS	ABS	ABS	ABS	1658	1792	2574	1
2	677GL	ABS	ABS	ABS	ABS	1126	1295	1327	1367	1503	1542	ABS	ABS	ABS	1665	ABS	1680	1810	3121	2
3	650GL	ABS	ABS	ABS	ABS	1060	1238	1266	1310	1450	1490	ABS	ABS	ABS	1617	ABS	1655	1795	3184	3

Lorain County

Well	Elev	TlGb	Mos	Lud	Skan	Mar	ChVl	Sen	MoNd	Edg	BoBl	Scho	CaCn	Esop	OrGl	Held	Rond	Sal	TD	Well
1	755GL	ABS	ABS	ABS	ABS	964	NI	1065	1090	1187	1228	ABS	ABS	ABS	ABS	ABS	1323	1353	2491	1
2	853GL	ABS	ABS	ABS	ABS	1156	NI	1257	1286			ABS	ABS	ABS					2653	2

Mahoning County

Well	Elev	TlGb	Mos	Lud	Skan	Mar	ChVl	Sen	MoNd	Edg	BoBl	Scho	CaCn	Esop	OrGl	Held	Rond	Sal	TD	Well
1	1060DF	2720	NI	ABS	NI	2730	2852	2860	2880	2972	3023	ABS	ABS	ABS	ABS	3200	3283	3383		1

Medina County

Well	Elev	TlGb	Mos	Lud	Skan	Mar	ChVl	Sen	MoNd	Edg	BoBl	Scho	CaCn	Esop	OrGl	Held	Rond	Sal	TD	Well
1	1250GL	ABS	ABS	ABS	ABS	1840	1960	1968	2004			ABS	ABS	ABS					5894	1
2	1075GL	ABS	ABS	ABS	ABS	1725	1837	1845	1875			ABS	ABS	ABS					3335	2
3								1740	1753	1860?	1893	ABS	ABS	ABS	ABS	ABS	1968	1987		3

Well	Elev	TlGb	Mos	Lud	Skan	Mar	ChVl	Sen	MoNd	Edg	BoBl	Scho	CaCn	Esop	OrGl	Held	Rond	Sal	TD	Well

Portage County

Well	Elev	TlGb	Mos	Lud	Skan	Mar	ChVl	Sen	MoNd	Edg	BoBl	Scho	CaCn	Esop	OrGl	Held	Rond	Sal	TD	Well
1								2477	2510	2634	2684	ABS	ABS	ABS	ABS	ABS	2810	2960		1
2	927	ABS	ABS	ABS	ABS	2252	2382	2395	2430	2553	2600	ABS	ABS	ABS	ABS	2733	2785	2882		2
3								2380	2413	2533	2575	ABS	ABS	ABS	ABS	ABS	2708	2870		3
4		ABS	ABS	ABS	ABS	2168	2317	2323	2353			ABS	ABS	ABS						4
5	1051GL	ABS	ABS	ABS	ABS	2125	NI	2275	2308			ABS	ABS	ABS					4216	5
6								2300	2326	2447	2500	ABS	ABS	ABS	ABS	ABS	2630	2800		6
7		ABS	ABS	ABS	ABS	2325	NI	2470	2505			ABS	ABS	ABS					4444	7
8	1160GL										2750?	ABS	ABS	ABS	ABS	2890	2960	3060	7089	8
9		2430	NI	ABS	NI	2434	2560	2570				ABS	ABS	ABS						9
10	959KB	ABS	ABS	ABS	ABS	2430	2550	2563	2592	2695	2748	ABS	ABS	ABS	ABS	2887	2936	3029		10
11							2800	2811	2827	2929	2985	ABS	ABS	ABS	ABS	3145	3227	3323		11

Stark County

Well	Elev	TlGb	Mos	Lud	Skan	Mar	ChVl	Sen	MoNd	Edg	BoBl	Scho	CaCn	Esop	OrGl	Held	Rond	Sal	TD	Well
1		ABS	ABS	ABS	ABS	3150	NI	3262	3275			ABS	ABS	ABS						1
2							NI	3340	3353	3427	3463	ABS	ABS	ABS	ABS	3647	3686	3776		2

Summit County

Well	Elev	TlGb	Mos	Lud	Skan	Mar	ChVl	Sen	MoNd	Edg	BoBl	Scho	CaCn	Esop	OrGl	Held	Rond	Sal	TD	Well
1		ABS	ABS	ABS	ABS	1625	1760	1767	1792			ABS	ABS	ABS					3447	1
2	1045GL	ABS	ABS	ABS	ABS	2089	NI	2197	2202	2315	2364	ABS	ABS	ABS	2453	ABS	2457	2551	2851	2
3								2320	2330	2458	2485	ABS	ABS	ABS	ABS	ABS	2575	2615		3

Trumbull County

Well	Elev	TlGb	Mos	Lud	Skan	Mar	ChVl	Sen	MoNd	Edg	BoBl	Scho	CaCn	Esop	OrGl	Held	Rond	Sal	TD	Well
1							2207	2215	2240	2360	2410	ABS	ABS	ABS	ABS	ABS	2518	2597		1
2	904KB	ABS	ABS	ABS	ABS	2181	2294	2310	2333	2435	2472	ABS	ABS	ABS	ABS	ABS	2580	2652		2
3	1141	ABS	ABS	ABS	ABS	2522	2647	2657	2678	2773	2816	ABS	ABS	ABS	ABS	ABS	2898	2966		3
4							3150	3158	3178	3240	3290	ABS	ABS	ABS	3380	3285	3400	3500		4
5	908GL	3123	NI	ABS	3133	3140	NI	3247	3263			ABS	ABS	ABS					5214	5
6	931GL	ABS	ABS	ABS	ABS	2690	2808	2812	2830			ABS	ABS	ABS					4931	6
7	992GL	ABS	ABS	ABS	ABS	2695	2813	2820	2838			ABS	ABS	ABS					4867	7

Tuscarawas County

Well	Elev	TlGb	Mos	Lud	Skan	Mar	ChVl	Sen	MoNd	Edg	BoBl	Scho	CaCn	Esop	OrGl	Held	Rond	Sal	TD	Well
1								2830	2835	2925?	2958	ABS	ABS	ABS	ABS	ABS	3042	3148		1
2								3151	3165	3252?	3272	ABS	ABS	ABS	ABS	ABS	3400	3495		2

Wayne County

Well	Elev	TlGb	Mos	Lud	Skan	Mar	ChVl	Sen	MoNd	Edg	BoBl	Scho	CaCn	Esop	OrGl	Held	Rond	Sal	TD	Well
1								2098	2114	2220?	2248	ABS	ABS	ABS	ABS	ABS	2319	2351		1
2								2402	2420	2520?	2555	ABS	ABS	ABS	ABS	ABS	2640	2663		2
3								2015	2030	NI	2150	ABS	ABS	ABS	ABS	ABS	2205	2225		3
4								2430	2447	2550?	2590	ABS	ABS	ABS	ABS	ABS	2695	2705		4

State of West Virginia

Hancock County

Well	Elev	TlGb	Mos	Lud	Skan	Mar	ChVl	Sen	MoNd	Edg	BoBl	Scho	CaCn	Esop	OrGl	Held	Rond	Sal	TD	Well
1	1052KB	4510	4515	4530	4565	4572	NI	ABS	4663	4669	4715	ABS	ABS	ABS	4877	4940	5080	5171		1

Province of Ontario

Elgin County

Well	Elev	TlGb	Mos	Lud	Skan	Mar	ChVl	Sen	MoNd	Edg	BoBl	Scho	CaCn	Esop	OrGl	Held	Rond	Sal	TD	Well
1	570GL	ABS	ABS	205	ABS	210	375	405	447	750	830	ABS	ABS	ABS	ABS	ABS	945	1055	2012	1
2	570GL						289	315	353	665	714	ABS	ABS	ABS	ABS	ABS	784	863	1900	2
3	570GL						170	204	237	518	564	ABS	ABS	ABS	ABS	ABS	631	700	1626	3
4	794						355	395	443	693	735	ABS	ABS	ABS	ABS	ABS	795	810	3763	4

5	573GL	ABS	ABS	ABS	ABS	565	711	725	763	893	956	ABS	ABS	ABS	ABS	ABS	1040	1063	2145	5
6	573GL	460	NI	ABS	NI	470	613	630	668	797	858	ABS	ABS	ABS	ABS	ABS	952	1004	1965	6
7	573GL	475	NI	ABS	NI	488	637	652	687	806	863	ABS	ABS	ABS	ABS	ABS	927	946	1952	7

Kent County

1	604GL	ABS	ABS	150	ABS	213	382	422	471	833	882	ABS	ABS	ABS	ABS	ABS	968	1133	3871	1
2	621						215	265	311	640	702	ABS	ABS	ABS	ABS	ABS	795	888	3876	2
3	608GL	ABS	ABS	278	ABS	297	460	491	536	840	905	ABS	ABS	ABS	ABS	ABS	1015	1127	1796	3
4	570GL	ABS	ABS	373	ABS	377	543	572	615	900	991	ABS	ABS	ABS	ABS	ABS	1104	1230	2273	4
5	573GL	ABS	ABS	316	ABS	323	500?	520	567	NI	895	ABS	ABS	ABS	ABS	ABS	1003	1106	2276	5
6	573GL	ABS	ABS	ABS	ABS	577	740?	762	813	1035	1100	ABS	ABS	ABS	ABS	ABS	1180	1198	2235	6
7	573GL	ABS	ABS	416	ABS	422	563	585	635	926	998	ABS	ABS	ABS	ABS	ABS	1087	1180	2256	7
8	573GL	ABS	ABS	ABS	ABS	290	435?	445	496	782	835	ABS	ABS	ABS	ABS	ABS	975	1150	2317	8
9	573GL	ABS	ABS	ABS	ABS	308	440?	454	505	782	837	ABS	ABS	ABS	ABS	ABS	975	1167	2268	9

Lambton County

1	657GL	ABS	ABS	195	ABS	255	380	422	450	830	873	ABS	ABS	ABS	ABS	ABS	937	1038	1683	1
2	677GL	ABS	ABS	125	ABS	215	345	395	435	860	907	ABS	ABS	ABS	ABS	ABS	995	1102	2220	2
3	684	ABS	ABS	128	ABS	223	357	427	503	935	976	ABS	ABS	ABS	ABS	ABS	1060	1173	4374	3

Middlesex County

1	707						220	255	296	652	693	ABS	ABS	ABS	ABS	ABS	770	825	3614	1

Norfolk County

1	570GL							412	448	585	622	ABS	ABS	ABS	ABS	ABS	700	724	1665	1
2	570GL	375	383	415	426	430	565	572	605	733	782	ABS	ABS	ABS	ABS	ABS	867	885	1789	2
3	570GL	606	613	626	636	639	783	788	820	934	991	ABS	ABS	ABS	ABS	ABS	1080	1107	2012	3
4	593DF	523?	NI	545	555	565	NI	707	NI	NI	900	ABS	ABS	ABS	935	ABS	950	1010?	4270	4

ERRATA

Due to a printing error, the lower left margin of Plate 4 is not completely reproduced, and several words are cropped. From top to bottom, please read "20 miles," "base of TICHENOR LS.," "base of CENTERFIELD LS.," "base of STAFFORD LS.," and "top of CHERRY VALLEY LS."

On the right-hand side of Plate 18, the designation "DEEP RUN SHALE" is incorrectly placed between the Borglum (Permit 4796) and Loree No. 2 (Permit 4797) wells. Although shown below the Centerfield Limestone, "DEEP RUN SHALE" should appear higher in the column between the underlying "base of TICHENORG LS." and the overlying "top of MENTETH LS."